AFRICAN HANDBOOKS: 3

The Food Resources of Africa

AFRICAN HANDBOOKS: 3

Edited by H. A. WIESCHHOFF

Committee on African Studies, University of Pennsylvania

THE FOOD RESOURCES OF AFRICA

By

Thomas S. Githens and Carroll E. Wood, Jr.

UNIVERSITY OF PENNSYLVANIA PRESS

THE UNIVERSITY MUSEUM

Philadelphia

1943

Manufactured in the United States of America

LONDON
HUMPHREY MILFORD
OXFORD UNIVERSITY PRESS

SECOND PRINTING, 1948

PREFACE

Africa, an important contributor to the world's agricultural supply, has now more than ever a vital share in the food economy of the world. The Japanese conquest of the Far East has deprived the United Nations of important food resources, and it is necessary to look for new fields to increase the production in those territories at the command of the United Nations. Africa's proximity to the major battle fronts makes it an area particularly strategic in location from which to supplement the depleted food supply, and the African natives contribute to this task in no small measure. Thus, the meager fat rations of Great Britain have come to depend largely on the amount of groundnuts, palm oil, and palm kernels which West Africa can furnish, and the Middle Eastern armies are supplied with food from East African territories.

The fact that, despite Africa's importance, only scant information about its food production is available makes the situation potentially dangerous. Not only might valuable food sources be ignored, but conversely Africa's productivity might be overestimated and lead to disastrous local famines if drained too much, as, indeed, occurred in some East African territories early this year.

For the United States, which contributes such a large share of food to both Allied and conquered nations, it is of special interest to learn and know about those food resources which now or in the near future may be available in other parts of the world.

These considerations prompted the decision of the Committee on African Studies at the University of Pennsylvania to undertake a program of Food Research. The present studies by Dr. Githens and Mr. Wood intend to give a general survey of Africa's agricultural resources, while in later volumes of the *African Handbooks* more detailed studies, both regarding specific areas and individual crops, will be forthcoming.

H. A. Wieschhoff

Philadelphia,
October 1943

CONTENTS

TROPICAL AFRICA: INTRODUCTION*

By

Thomas S. Githens

The character of the natural vegetation of Africa and of its agricultural products is determined by the fact that almost the entire continent lies within or close to the tropics, and differences in prevailing temperature due to latitude have therefore less influence than is the case on other continents. Of much greater importance agriculturally are—first, the amount of rainfall, which varies from over 60 inches a year in the rain forests to almost nothing in the deserts; and, second, the elevation above sea level, which has a determining influence on the prevailing temperature.

In addition to climatic considerations, there are certain other factors which have a most important influence on the practice of agriculture. One of the most important is the character of native farming. This is extremely primitive, even such simple farm implements as the plough being little used on the prevailing small native plots. Manuring to maintain fertility is practiced to a very limited extent. The usual procedure, at least in forested regions, is to cut down the trees over a small area, root out the stumps or allow them to rot, and then to dig up the ground with pointed sticks or similar primitive tools. Steel spades and hoes are often of crude type or even not available, and plowing with animals is almost unknown except in parts of North and South Africa. The ground is cultivated, usually with a succession of different crops, but often with only one, for five or six years until the soil is exhausted. The area is then allowed to revert to forest, and a fresh plot is similarly cleared and planted. After several years the forest will grow over the deserted plots; and with the lapse of years its fertility will be somewhat restored, and it can be cleared and used again. Only small plots around the village which are manured with house refuse are used year after year. In West and South Africa the use of the land is more continuous.

The practice of agriculture is also greatly influenced by the communal ownership of most of the arable land. The tribe or community holds title to all land, generally in the name of the chief; and he allots to each family the land which they will cultivate. Although the same plot may be assigned to the same family for several years, or until its reversion to forest, there is no personal

* Tropical Africa includes approximately the area between lat. 20° N. and lat. 20° S.

58
54
67
4
3
59
60
55
1
2
68
56
21
22
61
53
66
18
24
11
14
16
73
65
63
64
5
12
17
23
70
25
57
6
10
7
13
26
71
72
69
62
8
35
20
36
27
29
78
48
75
74
84
77
32
31
76
37
80
42
40
30
33
83
81
44
38
86
87
41
82
85
34
39
92
90
50
91
49
94
47
93
88
51
43
95
89
45
46
96

Fig. 1

List of Names for Numbers on Map

Aruwimi river	29
Ashanti	13
Atbar	67
Bahr el Ghazal province	62
Baro river	73
Bemba tribe	48
Benguela	47
Benguela plateau	43
Benue province	25
Benue river	24
Berber province	54
Bopora (6)	10
Borana district	63
Busoga, Lake Usoga	74
Cabinda province	44
Casamance	2
Cassinga	51
Catumbela	49
Congo river	27
Cuanza river	41
Dagomba tribe	12
Dar es Salaam	86
Darfur province	61
Dinka tribe	70
Djougou	14
Dongola province	59
Equatorville	32
Fort Jameson	94
Gambela	66
Gbanga	7
Gezira	56
Halfa province	58
Hargeisa district	64
Harrar	65
Ilorin	17
Kafue flats	96
Kamba tribe	78
Kaduna river	18
Kano province	22
Kasai province	38
Kasai river	42
Kassale province	55
Katanga province	39
Kikuyu tribe	77
Kisumu province	75
Kordofan province	60
Lake province	80
Leopoldville	30
Lindi province	85
Luangwa river	91
Lupweshi river	34
Makowa river	28
Mamprusi tribe	11
Maridi	69
Mayomba province	40
Monrovia	8
Mossamedes tribe	46
Mufindi district	81
Munshi tribe	26
Naaru river	72
Niger river	20
Novo Rendondo	50
Nuba Mt. Dar Nuba	53
Nupe tribe	16
Nyanza (Lake)	90
Nyanza province	76
Ovimbundu tribe	45
Railroad of Northern Rhodesia	95
Ruanda-Urundi	37
Rufiji river	87
Rungwe	82
St. Paul river	6
Senegal river	1
Sennar province	68
Shire highlands	88
Shire river	88
Stanley Falls	31
Stanleyville	31
Stephanieville	33
Tanganyika (Lake)	83
Temne tribe	5
Timbuktu	4
Tonj (Ibba) river	71
Tuareg tribe	3
Uele province	35
Uele river	35
Upoto district	36
Victoria Lake	84
Wasengi tribe	93
Wemba tribe	92
White Nile province	57
Yola	23
Zaria province	21
Zomba	89

ownership of most of the land except that given over to white proprietors or in the vicinity of white settlements.

In many, if not most, farming communities only the amount of food required by each family is kept by it. The excess is taken by the local chief and stored against famine or poor crops or sold for the benefit of the tribe or village. Again the personal incentive to grow larger crops or to make better use of the land is lacking. Inadequate storage and shipping facilities render the growing of excess crops useless.

The failure to use animal manures is due to the infrequency of mixed farming, by which is meant the keeping of livestock by the same persons who grow the crops. There are several reasons for this. One of the most important is the prevalence of trypanosomiasis, carried by the tsetse fly, over much of the most fertile part of Africa. Another reason is social or tribal tradition. The cattle and horse herders of North Africa consider farming beneath their dignity and prefer a nomadic habit of life to one of living in settled communities. Most of the land that is well suited to grazing has too low a rainfall to be productive of good crops. Many of the tribes south and east of the trypanosomiasis zone keep moderate flocks of sheep and goats as well as chickens, but although these may add some variety to the diet, the manure obtained is of little economic importance.

Owing to the difficulty of preparing the ground, it is a common practice to plant two crops, which mature at different times, interspersed with one another, and thus cereals and legumes, or maize and guineacorn, or yams and okra are planted together.

A factor which it is very difficult to evaluate is the influence of foreign settlement. The whole of Africa, with the unimportant exceptions of Liberia and Ethiopia, is today under the rule of European countries. Their governments have made relatively ineffective attempts to introduce among the native tribes the use of simple farm machinery, the practice of mixed farming, prevention of soil erosion, and other improved methods. Some colonies, however, established agricultural schools where natives are instructed by demonstration of improved method on government farms, but the results of these developments are still largely negligible. In many of the East and South African colonies, large grants of land are made to white settlers, who hire or induce the natives to cultivate the land under their direction and for their profit. The crops grown in land owned and managed by Europeans differ markedly in character from those grown by natives for their own use. There is more emphasis on crops raised for export and on foods not so well suited to African climatic conditions but which the whites desire because they have been accustomed to them at home.

The French colonial policy has encouraged the natives to consider themselves French citizens. The French colonial governors have introduced compulsory military service or labor on public works and have paid much attention to proper living conditions and nutrition in order to keep the workers in good condition. The British have left the native more to himself but by taxation have encouraged the growing of export crops, which can be sold to obtain money for taxes. The Germans and Belgians formerly used forced labor amounting almost to slavery, but the position of the natives is now much improved, both in the Belgian Congo and in the former German colonies of East Africa.

Owing to the native practice of cultivating small plots of ground and the poor facilities for obtaining precise information in the scattered villages and difficulty of travel, it is virtually impossible to determine with any degree of accuracy the total acreage of any crop under native cultivation, or the total yield. Such figures as are given are largely estimates based on partial observation.[1] For this reason, it is not uncommon for two well-informed observers to cite figures for a given crop, one of which will be several times more than the other. Even wider differences are seen when successive years are compared. For example, the rice crop in Tanganyika is said to have been 6,513 tons in 1936 and 73,675 tons, or 10 times as much, in 1937. The crop of groundnuts in Kenya is given as from 600 to 2,200 tons between 1937 and 1939.

A study of the figures gives rise to the belief that, in cases of wide discrepancy, the lower figure is more likely to be based on incomplete reports than is the higher to represent exaggeration. For this reason, and because this report is largely an attempt to evaluate the potential capacity of African agriculture as based on published reports, the largest figures have been included in the tables in the following chapters except when they were out of line with all other years and other sources of information. Very few figures are available for 1941 or subsequent years and reports earlier than 1935 have almost never been included, almost all the statistical data being for the years 1935 to 1940 inclusive.

[1] The fields are not harvested in the European sense. A fruit or root is taken without disturbing the whole plant. The plantings are mixed in age and kind of plant, and the owner picks what he needs for each meal. It would be impossible to measure by any statistical method based on harvested or sold crop, the total amount or value of agricultural production. Census figures are only approximations or even guesses. Excerpted from H. L. Shantz, "Agricultural Regions of Africa," *Ec. Geog.* XVI (1940) 11.

From the standpoints of rainfall and elevation, Africa may be roughly divided into the following regions: 1. The Western Mediterranean Littoral, including the Atlas Mountains; 2. The Desert Region, with a rainfall of less than 15 inches; 3. Rain Forest and Park Savannah, with a rainfall of over 40 inches yearly and elevation of 600 to 3,000 feet; 4. High Grass Savannah with a similar elevation and a rainfall of 15 to 40 inches; 5. Mountain and High Plateau with elevation above 3,000 feet and variable rainfall.

MEDITERRANEAN LITTORAL

This area includes Morocco, Algeria, and Tunisia (with the exception of the large desert areas in the southern part of each country) and is for the most part characterized by the presence of the Atlas ranges. The climate is of the Mediterranean type with rainy winters and dry summers, being thus similar to that of southern Europe where agricultural conditions are likewise much the same. The most important food crops, both from the standpoint of area and that of production, are the cereals—hard and soft wheats, barley being grown for the most part. Smaller amounts of oats, maize, sorghum, and rye are also grown. Fruit trees and vines, including olives, figs, dates, oranges, lemons, tangerines, and other citrus fruits; grapes and raisins; peaches, almonds, and other stone fruits are quite important in the region. Root crops are less important than in other parts of Africa, but potatoes and other root crops are raised to some extent, as well as a variety of truck crops including beans, tomatoes, artichokes, peas, lettuce, and others in small amounts. Food production varies greatly according to weather conditions and in some years is insufficient for the needs of the local populations.

THE DESERT REGION

The great desert of Africa extends across the entire continent above lat. 15° N., from Mauritania on the west coast to Egypt on the east, and also includes the Somali peninsula. It comprises most of French West Africa, southern Algeria, Tripoli, Egypt, the Anglo-Egyptian Sudan, the Somalilands, and a large part of Ethiopia. Throughout the greater part of its extent there is no agriculture, the land being entirely barren or used for grazing only. Here and there, scattered throughout its extent, are small oases where figs, dates, olives, and other crops are grown. Toward its eastern end it is cut by the Nile, along which for many miles irrigation makes it possible to raise a wide variety of crops, including cereals such as wheat, oats, barley, rice, maize, and sorghum; root crops including white and sweet potatoes, yams, beets, and turnips; vegetables, beans and peas, tomatoes, eggplant, pumpkins, and squashes, as well as sugar cane. Similar conditions obtain along the river valleys of Ethiopia and Somaliland. Where there is occasional rain, cereals suited to dry conditions are cultivated, including sorghum in the wetter parts and bulrush millet, digitaria, and pennisetum in the dryer. Cotton and sesame are grown especially along the Nile.

In all of this region animal husbandry, including the raising of cattle, sheep, and goats, is more important than the growing of plant crops.

RAIN FOREST AND PARK SAVANNAH

Agriculturally the tropical Rain Forest is probably the most important region of Africa. Many of the indigenous trees are of great economic importance, and the abundance of moisture and the tropic heat make the growth of planted crops rapid and profuse. The typical rain forest is limited to the western half of Africa, including the Guinea coast and the central basin of the Congo River. Around it extends a belt of Park Savannah, characterized by the presence of large stands of trees interspersed with rich meadowland. The rainfall is sufficient for almost all crops, and it is little less productive than the rain forest. The Rain Forest and Park Savannah, which have a rainfall over 40 inches yearly and an elevation above sea level of 600 to 3,000 feet, include all the countries of the west coast (the southern part of Senegal, Gambia, French and Portuguese Guinea, Sierra Leone, Liberia, Ivory Coast, Gold Coast, Togo, Dahomey, Nigeria, Cameroon) and those of the adjacent West Coast of Equatorial Africa (Gabon, Middle Congo, Belgian Congo, and the northern half of Angola).

Throughout this area the cultivation is almost entirely native, there being much less land allotted to white settlers than in East and South Africa. In Nigeria, Gold Coast, and Sierra Leone, almost the only land alotted for white cultivation is that used for experimental gardens; in the Belgian Congo about 6 per cent of the cultivated land is in white plantations; and in the French colonies the amount of land in white hands is less than this. In Liberia, the only white plantations are those of rubber trees held by the Firestone Tire Company.

Agricultural conditions in all the colonies of the West coast are esentially similar, although there are differences in the relative importance of the various crops in the several colonies owing partly to local preference for certain foods and partly to pressure from administrators to grow special crops for export. In all the colonies there is a pronounced distinction between the crops grown for local consumption and those grown principally for export. The white administrators have encouraged the latter, as they are the most important source of revenue. Because of the greater interest of the French colonial officials in native nutrition, statistics in regard to the total yield of each important crop are prepared regularly. In the British colonies the only records, as a rule, are those of exports, and no statistics are prepared on the crops used as food by the natives.

The most important export crops of the Rain Forest-Park Savannah region are the fruits of certain trees which grow near the coast. Greatest in volume is the fruit of the oil palm of which 700,000 tons of nuts and 200,000 tons of oil are exported in a single year. The oil is used for margarine and for soap manufacture. A very large but undetermined amount of the oil is used locally by the natives for cooking. Even greater in value is the crop of cocoa beans. All of this, amounting to about 500,000 tons yearly—more than half the world's supply—is exported. Bananas are cultivated, especially in the British colonies, and over 100,000 tons were exported in 1940. Of great importance also are coconuts, of which as much as 10,000 tons have been exported in a single year, mostly from Dahomey and Belgian Congo, as well as 180 tons of coconut oil and several thousand tons of copra.

More or less successful attempts have been made in recent years to introduce the cultivation of citrus fruits in the coastal areas of the British colonies. Many oranges are small and of the bitter variety and cannot compete with those of southern Europe, but 349 tons were exported from Nigeria to other colonies in 1938. Limes have been grown for several years in the Gold Coast; but owing to poor storage and refrigeration facilities, the fruit tends to spoil and greater success has been attained with the export of prepared lime juice. In one year 45 tons of fruit and 335,000 gallons of juice were exported, and a 600,000-gallon export was anticipated for 1942.

Kola nuts are among the more important of the minor crops in several colonies, almost 10,000 tons being exported annually for use in caffeine drinks. Coffee is grown less on the Guinea coast than in West Africa, but 35,000 tons are exported from Belgian Congo and Angola. Apart from peanuts, which are cultivated more extensively farther inland, the only other export crop is ginger, of which over 2,500 tons are exported, chiefly from Sierra Leone.

In addition to these products, several crops more characteristic of High Grass Savannah are also raised in the Rain Forest belt.

The crops for local consumption are chiefly cereal grasses and starchy roots. Of the cereals, maize is the chief crop near the coast and in Congo and Angola. As we go inland this is gradually replaced by guineacorn, which gives place to bulrush millet and then to other millets which withstand drought. Rice is cultivated along the rivers and in wet ground. Of the root crops, cassava is grown throughout but is exceeded by yams in the coastal zone. Sweet potatoes and coco-yams become increasingly important in the drier parts inland. White potatoes are cultivated extensively in Belgian Congo. Many legumes are raised in small plots, especially haricot beans in the French colonies and Angola, and cowpeas and pigeon peas in the English. Peanuts are widely grown at a little distance from the coast. Many other vegetables and fruits are used to diversify the diet, including okra, which forms part of every garden, and many cucurbits such as pumpkins, squash, watermelons, muskmelon, and gourds. Plantains largely replace bananas in the native dietary. Eggplants and tomatoes are universally grown. Red peppers and chillies are used everywhere as seasoning. Pineapples are becoming more popular and, although most of the crop is consumed locally, over 500 tons are exported annually.

HIGH GRASS SAVANNAH

This region extends in a belt around the Park Savannah belt. The rainfall is from 20 to 40 inches, and it is adapted for both agriculture and grazing. In this belt, agricultural practices are much the same as in the Rain Forest belt, but there is a difference in the crops. The oil trees, which contribute products for export near the coast, are not found inland, and the distinction between export and native food crops is less sharp. The most important export crops are oil-bearing seeds, and the amount of these consumed locally approximates or exceeds that exported. In this category, peanuts are the most important, annual export from West Africa being over 850,000 tons. Minor exports include benniseed (about 27,000 tons) and shea tree nuts (approximately 10,000 tons). The oil or nuts from all of these are much used by the natives and form the chief source of fat in the diet. The main cereal crop is bulrush millet, although guineacorn is planted in the wetter parts of the zone and other millets in the drier parts. Root crops include sweet potatoes, coco-yams, and cassava. Most of the belt is too dry for true yams. The same vegetables are grown as in the Forest belt. There are few fruits, but watermelons are more extensively used than near the coast.

MOUNTAIN AND HIGH PLATEAU

This type of country, mostly more than 3,000 feet above sea level, occurs in British East Africa, except for a narrow strip along the coast, and also includes most of Africa below lat. 10° S. This area is much more suited to white settlement on account of the cooler climate, and a larger proportion of the ground is under white cultivation. Climatic conditions are more like those of Europe, and most of the vegetables and fruits grown in England have been successfully introduced. The freedom from trypanosomiasis makes mixed farming practical, and for this reason crops are much more diversified. Mountains and valleys and lakes are interspersed, and the country cannot be divided into zones as is possible in west Africa. For this reason generalizations are not readily made. Export crops are mainly grown on land owned or controlled by whites, and the natives depend for money on wages as well as on "money crops." The chief exports from British East Africa are coffee and tea; oil plants, including peanuts, benniseed, and coconuts; and cereals, such as rice, maize, and wheat. Small amounts of chillies and sugar are also exported. The native crops include many kinds of cereals, cassava, sweet and white potatoes, peanuts and other legumes, pumpkins and melons, and, near the coast, sugar cane and bananas.

INDIVIDUAL CROPS

For the consideration of individual crops, a somewhat different grouping of the regions of Africa is more convenient as many plants extend over several zones; and a regional rather than a meteorological grouping gives a better idea of the general crop picture. From this standpoint, Africa may be divided in the following areas:

1. Mediterranean Littoral. Including Morocco, Algeria, Tunisia, Tripoli, and Egypt. (See pp. 55ff.)

2. West Africa. Including French West Africa (Senegal, Mauritania, French Sudan, French Niger, Ivory Coast, French Guinea, Dahomey), Gambia, Portuguese Guinea, Sierra Leone, Liberia, Gold Coast, Togo, Nigeria, Cameroon, and French Equatorial Africa (Chad, Ubangi-Shari, Gabon, Middle Congo). French Equatorial Africa forms a link with West Equatorial Africa and is not always classed as part of West Africa. Its northern provinces are, however, related agriculturally to those of French West Africa and, as most agricultural statistics are given for the entire colony, it is not possible to separate Gabon and Middle Congo from them in the discussion of crops.

3. West Equatorial Africa. Including Belgian Congo and Angola.

4. North East Africa. Including Anglo-Egyptian Sudan, Ethiopia, and the Somalilands.

5. East Africa. Including Uganda, Kenya, Tanganyika, Nyasaland, and Northern Rhodesia.

6. South Africa. Including South West Africa, Bechuanaland, Southern Rhodesia, Mozambique, and the Union of South Africa.

(*Note:* The last of the above areas is not treated in the present book, nor are the islands off the coast, such as Madagascar and Zanzibar.)

Chapter I

WEST AFRICA

The native diet in the region of the Guinea coast consists chiefly of cereals and starchy roots Proteins are supplied by various legumes, and fats by oil palm, benniseed, shea butter, and peanuts. The only fruits eaten in large amount are bananas and plantains. Cola is used considerably as a stimulant, and several peppers serve to add flavor to the food.

CEREAL CROPS

There are four important cereal grains grown in West Africa and several others which are grown in smaller amount. The particular grain cultivated in each region depends chiefly on rainfall or water supply. In the wettest regions near the coast and along rivers, rice is raised rather extensively but much less than the other three. Where there is considerable rainfall in the Tropical Forest belt, maize is the chief cereal. Farther from the coast, extending from the Forest across to Park Savannah belt, guineacorn is grown most extensively. Farther inland, where the rainfall is less, and extending in places to the very edge of the desert, bulrush millet replaces the other cereals. There is no sharp line dividing these belts, and in all but the very driest and wettest parts all three are cultivated side by side or in the same field; but as one travels into the interior the area alloted to each alters as indicated.

Certain smaller millets are cultivated with bulrush millet or alone on the edge of the desert where it is too dry for guineacorn.

A few cereals more recently introduced are grown often locally, especially where white influence is strong. These are wheat, barley, and oats. They are raised chiefly in places where the desert is interspersed by wet areas, as along the Niger River and in the oases.

All of these cereals, with the exception of certain of the millets, are of foreign origin, having been introduced into Africa by European explorers within the last four hundred years. It is of interest, as showing the adaptability of the natives, that all of them are now grown by even the less advanced tribes.

GUINEACORN: Kaffir Corn, Durra, Dawa, Napira, Mtama, Great Millet. This is probably the most important native African cereal. It is a tall grass, 4 to 8 feet high, with a spike often several feet long.

The grain is small but abundant, yielding up to 1,200 pounds per acre, although the average is not much more than half this. It grows best with a rainfall of from 20 to 50 inches. There are a number of varieties. Two of the more important being *Sorghum vulgare guineense*, which grows where the rainfall is that mentioned, and *S. v. cernuum* (maskwari) suited to drier regions. The varieties durra and kaffir corn are rarely grown in West Africa. One variety known as "early" is planted in May and is harvested in November. It is often planted with maize or bulrush millet which is gathered in August before the guineacorn has reached its full growth. A late variety planted at the same time does not ripen until late December. This variety is often planted interspersed with peanuts. The grain of *S. v. guineense* makes poor flour for breadstuffs but is used as a porridge. *S. v. cernuum* may be used for bread. The grains are divided according to color, as white (fura-fura), yellow (kauri), and red (used as fodder).

Other varieties of guineacorn are also extensively grown. One (*Sorghum vulgare saccharatum*) has much sugar in the stems and is grown as a substitute for sugar cane. Varieties known as *halepense* and *sudanense* are grown for cattle fodder and *Sorghum vulgare technicus* for broomstraw.

Guineacorn is the chief cereal crop in Senegal and Guinea, in the northern part of Ivory Coast, Gold Coast, Dahomey, and Nigeria, and in the wetter parts of Mauritania, French Sudan, and Niger. Nearer the coast it is gradually replaced by maize and toward the north by bulrush millet.

The following table shows the total crop of guineacorn in a favorable year as far as figures are available:

		Yield per acre			
	Acres	*Max.*	*Min.*	*Average*	*Crop in tons*
Cameroon					150,000 (all millets)
Dahomey	4,500,000			(900) lbs.*	2,000,000 (all millets)
Ivory Coast	625,000			(800)	246,000
French Togo					20,000 (all millets)
Gold Coast		1,005	235	750	
Nigeria		1,300	212	1,000	

* All yields in parentheses in this and later tables are calculated by dividing the crop weight by the number of acres.

Bulrush Millet: Pencillaria, Gero, Machewere, Dokhn. Bulrush millet is a tall grass, but not as tall as guineacorn, reaching a maximum of 10 feet, with spikes a foot long but heavier than those of guineacorn. It flourishes, however, in drier regions where the rainfall is about 15 to 10 inches yearly or even less. Where the rainfall is light, it is the only cereal grown on a large scale—as near the Sahara Desert. It is the chief grain crop of Senegal, Mauritania, French Sudan, Niger, and Chad, and of the drier parts of Guinea, Ivory Coast, Dahomey, Nigeria, and Cameroon. It requires little cultivation. Although the yield of grain per plant is greater than that of guineacorn, it is not planted as close and the yield per acre is about the same. From every pound of seed sown, a crop of 300 to 400 pounds of grain is obtained, which is much larger than the yield of most other cereals. It ripens earlier than guineacorn. There are two varieties: early millet, which grows very rapidly, is planted in early April and harvested in late June; the late variety is planted at the end of May and ripens in December. This variety is often interplanted with guineacorn, yams, peanuts, bambarra nuts, or cucurbits. The yield per acre is fairly constant, from 800 to 1,000 pounds of heads per acre, yielding 400 to 600 pounds of threshed grain. It makes better bread than does guineacorn.

The following table shows the statistical data for 1936:

	Acres	*Yield per acre*	*Crop in tons*	
Senegal	2,300,000	(550) lbs.	633,200	Figures may include other millets
Mauritania			37,000	
French Sudan			655,300	
Niger			30,000	
Ivory Coast	1,115,000	(320)	176,000	
Dahomey	6,480,000	(625)	2,000,000	
French Togo			20,000	
Cameroon	1,110,000	(440)	290,000	

Fonio: Late Millet. This cereal requires very little water and is grown in the nearly-desert regions along the edge of the Sahara. It is an important crop in the French Sudan and Niger but grown less in the drier parts of other colonies. In the northern part of Nigeria it is raised for fodder rather than for human food, but the northern tribes of Gold Coast and of Guinea use it as a staple of the diet. The yield is about 400 pounds per acre.

The following table shows the tons harvested in 1937:

French Guinea	160,000
Ivory Coast	19,300
French Sudan	4,000

Finger Millet: Acha, Hungry Rice; Iburu. These are grown occasionally by certain tribes, e.g., by the Dagomba of the northern Gold Coast, the pagan as well as Hausa tribes in poor sandy soil in the higher and drier parts of Nigeria; by certain tribes in Gambia, in Temne province of Sierra Leone; and in the loop of the Niger River. The plants are much smaller than those of the other millets and the yield of grain is low, averaging about 150 pounds per acre.

A similar millet, Tamba, is used in Nigeria for making beer and is eaten only in famine years. It is cultivated to a greater extent in the drier parts of East Africa.

Another millet, *Panicum burgii*, is grown in the Niger Territory.

MAIZE: Corn, Indian Corn. Maize is one of the two chief cereals of West Africa, the other being guineacorn. Maize grows where the rainfall is from 30 to 60 inches, guineacorn where it is from 15 to 50 inches, so that in the greater part of their range the two can be planted side by side. Maize is generally planted in late April on ridges six feet apart, with guineacorn between. The maize is harvested in August before the guineacorn is ripe. More rarely, maize is planted with the late rains in August and the ears gathered in January. In the coastal colonies, it is the chief or only cereal crop for a distance of 100 to 200 miles from the coast. In Dahomey it is the chief cereal, and in Sierra Leone second only to rice. In Senegal, Mauritania, French Sudan, and Niger it is grown only along the rivers. In Gabon it is the chief cereal sold in the local markets.

Both sweet corn and hard corn are extensively grown, the former being eaten fresh, the latter ground and made into porridge or cakes. Both red and white types are seen. A native variety of sweet corn with dark grains is considered most palatable. The yield is usually about 1,000 pounds of grain per acre, but varies from as little as 300 pounds on plots planted for several years without manure, to over 2,000 pounds under favorable conditions and in fresh ground.

	Acres	*Lbs. per acre*	*Crop in tons*
French Equatorial	92,000	(830)	38,000
Mauritania	45,000	(400)	9,000
French Sudan	300,000	(500)	74,400
Niger	90,000	(460)	21,000
Senegal	48,000	(770)	18,500
French Guinea	160,000	(700)	55,000
Sierra Leone	6,200	1,000	800
Ivory Coast	625,000	(470)	147,000
Gold Coast		600–2,300	
Dahomey	560,000		180,000
Nigeria		1,000–2,000	Exp. 20,000
Cameroon	150,000		36,000

RICE: Several varieties of rice are grown in West Africa. There are two general types: swamp rice which is raised along rivers and in mangrove swamps, and which is inundated when planted; and upland rice which is not planted near water and is rarely flooded but which requires an annual rainfall of 30 to 50 inches. Most of the rice grown in West Africa is of a red variety not suited for export but preferred by the natives. It is not cultivated carefully as it is in the Far East, and the yields are smaller. It is also carelessly harvested and much remains ungarnered. In most of the colonies, rice is a secondary crop; only in Sierra Leone is it the chief cereal. It forms an important element of the diet in Gambia, Guinea, Senegal, and French Sudan. In Nigeria it is grown by the Nupe and Ilorin tribes along the Kaduna and Baku rivers, and about 12,500 acres are planted in the Niger delta. In Gold Coast it is limited to a small strip along the coast. In Dahomey it is cultivated at Djougou in the north; and in Senegal, in Casamanca province, and along the Senegal River. In Niger, swamp rice is raised along the loop of the Niger River. In Liberia much is grown along St. Paul's River.

	Acres	*Lbs. per acre*	*Crop in tons*
Mauritania	1,750	(600)	500
French Sudan	250,000	(600)	73,000
French Equatorial	1,500	(900)	7,000
Niger	8,000	(400)	1,600
Senegal	145,000	(700)	55,000
French Guinea	850,000	(500)	205,600
Sierra Leone	300,000	(1,200)	210,000
Ivory Coast	460,000	(400)	878,000
Gold Coast	230,000	690	
Dahomey	1,100	(400)	200
Nigeria		400–4,000	40,000
Cameroon	250		50

Only in Sierra Leone is rice an important export; 6,000 tons were exported in 1940.

WHEAT: Wheat is not an important crop anywhere in West Africa. The climate is not suitable, and it can be grown only in the few places which are fairly well watered and not excessively hot, north of 11°. The native method of growing small plots leads to small crops in any one region. In Mauritania some wheat is grown in the oases, and in 1937 the crop was estimated as 40 tons. In the Sudan it is raised in the Tuareg grasslands and near Timbuctoo. In Niger it is grown in isolated spots in the Niger Valley; and in 1937, 800 tons were harvested. Small amounts are raised in Zaria and Kano provinces in northern Nigeria. It is eaten mainly by the white population or sold to other colonies. Little or none is exported.

BARLEY: Barley is grown only at a distance from the coast, in much the same regions as wheat. In Mauritania 90 tons, and in the Niger 100 tons, were harvested in 1937. In French Sudan it is cultivated to a small extent in the Tuareg grasslands. Little if any is exported.

LEGUMES

Peanuts are by far the most important legume crop of West Africa but differ from the other legumes in being a source of oil rather than of protein. A large number of other legumes are grown throughout West Africa; but except for the cultivation of haricot beans in the French colonies under white influence, they are grown in small plots around villages, and there are no figures available to indicate the actual crop or the acreage utilized. The following are the kinds most commonly grown.

HARICOT BEANS (Navy Beans) are grown extensively in all the French colonies and large crops are gathered, a small part of which is exported. They are more often grown alone than other legumes, which are almost always planted interspersed with cereals or root crops. The following shows acreage and yield in 1937:

Colony	*Acreage*	*Crops in tons*
Mauritania	50,000	10,000
French Sudan		45,000
Senegal	135,000	27,000
Dahomey	75,000	4,500
Ivory Coast	230,000	19,300
Cameroon	25,000	300
French Equatorial	25,000	5,000

LIMA BEANS. A small variety known as "popondo beans" are grown everywhere along the Guinea coast, especially in Nigeria and Liberia, in the wetter parts. They are planted in August or September, interplanted with other crops. The maximum yield is about 150 pounds per acre. No figures are available for the crop.

COW PEAS: Dagarti Beans. Catjang Peas. These are favorite vegetables in all tropical Africa and are found in most gardens. They are grown throughout Nigeria and Gold Coast and are one of the chief foods of the Mamprusi and Dagomba in northern Gold Coast. They are planted in May, usually mixed with guineacorn, and harvested in November. The yield is about 80 pounds per acre.

PIGEON PEAS: Congo Beans. These are grown where the rainfall is less than that best suited to cowpeas and are widely cultivated in the drier parts of all the West coast colonies. They are eaten either as the green pods or the ripe seeds, and are a favorite in Guinea, Sierra Leone, Dahomey, and Ivory Coast. In Nigeria they are grown chiefly in Zaria province, generally interplanted with cereal or root crops, but in northern Gold Coast they are planted as hedges in late June and harvested from December to February.

LAB LAB BEANS. These are grown everywhere as a cover crop and used as food, either as green pods or ripe seeds, chiefly in Kano Province in northern Nigeria and in the loop of the Niger River. The yield of dry seeds is 250 to 400 pounds per acre.

SWORD BEANS. Sword beans are raised in small amount throughout tropical Africa by the pagan tribes. They are favorites in northern Nigeria and near Gbanga in Liberia.

LOCUST BEANS. These are large edible beans which grow on a native tree. They are cultivated extensively but not in large stands. Most of the trees are wild, and many are found in Dahomey. The pulp is eaten, and the seeds are roasted and used as coffee.

BAMBARA GROUNDNUTS. This plant has an underground nut something like the peanut. It is widely grown in the Savannah belt and is a staple food in north Gold Coast, northern Nigeria, and in the loop of the Niger River. It is planted in late June, with yams or cereals or alone, and is harvested in November. The yield is about 500 pounds per acre. Seven thousand two hundred tons were harvested in Niger in 1937.

PEANUTS: Groundnuts, Gedda. The peanut is perhaps the most universally cultivated crop in Africa. It is the chief source of fat or oil in several of the colonies and one of the most important export crops in all of them. As a source of fat, its only rival is the oil palm, and this grows only near the coast in the Rain Forest. The peanut is often planted alone but more often interspersed with a cereal or root crop. It is one of the chief crops in the belt known as the Sahel, which extends from the 9-inch to the 25-inch-rainfall isohyet, but is grown up to the 50-inch isohyet. The only colonies in which peanuts are not widely grown are Sierra Leone and Liberia, but they are less important in the coastal zone of each colony than farther inland. Thus they are found less frequently in southern than in northern Nigeria and less in Gold Coast proper than in Ashanti. In Senegal and Gambia, except near rivers, they are the chief native food. They may be planted with the early or late rains and harvested in three to five months. In Ashanti they are generally planted in May as a single crop, or with a cereal grain, and are harvested from August to November. In northern Nigeria peanuts are often used as the last crop before the ground is allowed to revert to forest. It is a common experience that they do not exhaust the soil, and can be planted repeatedly on the same plot without substantial loss of yield.

Although properly classed as a legume, they differ from the peas, beans, and pulses in being a source of fat in the diet as well as of protein and carbohydrates.

The table shows the yield for a good year:

Colony	*Acres planted*	*Lbs. per acre*	*Crop in tons*	*Export in tons*
Mauritania	6,500	(400)	1,300	
French Sudan	275,000	(750)	104,000	900
Niger	475,000	(315)	75,000	
French Equatorial	110,000	(300)	17,000	6,700
Senegal	1,260,000	(200)	660,000	508,000
Gambia				76,000
French Guinea	130,000	(560)	36,600	2,000
Sierra Leone				5,900
Ivory Coast	350,000	(230)	40,000	
Gold Coast		700		
Dahomey	20,000	(300)	3,000	4
Nigeria		500–1,500	473,000	265,000
Cameroon	350,000	(230)	40,000	12,000
Portuguese Guinea				31,500

ROOT CROPS

YAMS. There are several species grown in West Africa, and in most references these are reported together so that there are no available figures as to the amount of each raised in any locality. The chief types are: 1. White Yam. This seems to be the favorite in most localities. The tubers weigh several pounds and keep well in storage. 2. Yellow Yam. Has similar tubers, but they do not keep well. 3. Water Yam. This has the largest tubers, weighing up to sixty pounds. Other species which are less often planted are: 4. Chinese Yam, which grows in drier ground and has small tubers, like sweet potatoes. 5. Bulb-bearing Yam, which bears bulbs the size of apples in the axil of the leaves. 6. Three-leaved Yam. The tubers contain an acrid substance which is destroyed or dissipated by cooking.

The white yam grows best with a rainfall of 30 to 50 inches, in the same belt as guineacorn, and through this belt yams are the main root crop. They are planted from January to April, and the tubers are harvested in about ten months, from November to February. Where rotation of crops is practised, yams are often planted as the first crop in new ground, interplanted with some more rapidly maturing crop such as beans, okra, pumpkins, or watermelon. Fra-Fra potatoes, which mature in six months, are sometimes planted in the same field in June, and the two dug up together in December.

Most of the crop is consumed locally, but there is a large internal trade from areas where it is grown extensively to other regions or colonies. Yams are widely cultivated in all the coastal colonies, but in French West Africa it is too dry except along the Niger River and near Lake Chad. The yield per acre may run as high as 12,000 pounds, but is usually not more than 2,500 to 3,000 pounds. In Gabon a great deal is sold in the local markets.

There are no figures available for the annual harvest except for certain French colonies which gave the following for 1937–38:

Colony	*Acres*	*Lbs. per acre*	*Crop in tons*
Niger	12,500	(4,160)	26,000
French Guinea	225,000	(88)	11,000
Ivory Coast	450,000	(2,750)	618,000
Dahomey	95,000	(6,300)	300,000
Cameroon	50,000	(4,000)	100,000

COCO-YAMS: Eddo, Taro, Caladium. The coco-yam was introduced into Africa by Portuguese travelers several centuries ago, and its cultivation has become widespread in West Africa, although it is planted less frequently than true yams and cassava. It will grow where it is too wet for yams and does best in wet, shady ground. Coco-yams are planted among the cocoa trees near the coast and farther inland in groves of the locust bean.

In many places it is being replaced by the New or False Coco-Yam (tania), which is now preferred by the natives of Gold Coast and Sierra Leone. Coco-yams are not exported but are consumed locally, and crop figures are available from only a few colonies. No distinction is made between the two species. The following figures are for 1937–38:

Colony	*Acreage*	*Lbs. per acre*	*Crop in tons*
French Guinea	1,500	(800)	600
Ivory Coast	25,000	(3,800)	47,600
Cameroon	125,000	(4,000)	250,000

CASSAVA: Manioc, Rogo. There are two species, Bitter Cassava, which contains hydrocyanic acid and which must be ground and thoroughly washed before eating, and Sweet Cassava, which contains no hydrocyanic acid and is not poisonous. Tapioca is a granulated starch prepared from bitter cassava by boiling the ground roots in water, straining, and drying on hot stones. This is the usual procedure in America, but tapioca is rarely prepared in West Africa except in Togoland. Two varieties of bitter cassava are recognized, the white and the red. The root of sweet cassava is darker in color and it is also called black cassava. Bitter cassava is a native of tropical America. Sweet cassava is said to be indigenous.

The three types are all grown extensively, and in agricultural statistics no distinction is made between them. Cassava may be grown under widely varying conditions and is universally cultivated from the Rain Forest almost to the edge of the Desert. One or more types of cassava are cultivated in all the colonies of West Africa and in Liberia. The method of planting and harvesting varies greatly in the different tribes. Cassava is a long-lived perennial and, if only part of the roots are harvested each year, harvesting may be continued for four or five years from one planting. Where ground is allowed to revert to forest after a few years of cultivation, planting is sometimes done the year before reversion and roots gathered while the plot is returning to the wild for several years. In other cases, planting is made in the second year of rotation, crops taken each year, and the last roots dug up just before the plot is deserted. The roots grow large enough to harvest in eight months; and the Dagomba of northern Gold Coast plant it in May, on ridges with yams between, and harvest it in January. In Niger, it is planted in February and dug up in October. In Gold Coast, experiments showed that the highest yields were obtained if the roots were collected fifteen months after planting. According to Shantz and Marbut, the roots require four years to reach their full size, and then harvesting may continue for four years more. The yield per acre is quite variable but may amount to over 10 tons an acre.

The following figures are mostly from 1937:

Colony	*Acreage*	*Lbs. per acre*	*Crop in tons*
French West Africa	900,000	(2,200)	1,000,000
French Equatorial	150,000	(10)	800
French Sudan	75,000	(2,800)	107,800
Niger	75,000	2,400	90,000
Senegal	87,000	3,200	140,000
French Guinea	125,000	1,600	100,800
Ivory Coast	420,000	10,000 (400)	80,000
Gold Coast		9,000 to 29,000	
Dahomey			300,000
Nigeria		6,000 to 20,000	
Cameroon	250,000	800	100,000

From Togo, over 2,000 tons of tapioca were exported in 1938.

Sweet Potatoes are grown everywhere in tropical Africa where the rainfall is 30 to 50 inches, or even less than this where water is available. In most colonies yams or cassava are grown in greater quantities, but sweet potatoes are little less important. In the Gold Coast they are planted alone in June and harvested in December. Much more is grown in the northern parts of Nigeria and Gold Coast than in the southern. They are cultivated more extensively in Liberia, Ivory Coast, Guinea, Dahomey, and Cameroon than in the other coastal colonies. Much is sold in the local markets in Gabon.

The following figures are mostly for 1937:

Colony	*Acreage*	*Lbs. per acre*	*Crop in tons*
Mauritania	1,250	(1,280)	800
French Sudan	55,000	(2,350)	64,000
Niger	3,000	3,600	5,400
French Guinea	61,835		45,700
Ivory Coast	87,000		75,300
Gold Coast		5,500	
Dahomey	17,500		21,000
Nigeria		2,000–10,000	
Cameroon			300,000

Fra-Fra Potatoes: Fura-Fura Potatoes, Fabirama, Ḍaso. This indigenous crop is a favorite food in all parts of West Africa, but grown less often and in smaller plots than yams or cassava. It is raised more in northern Nigeria and northern Gold Coast than in the other coastal colonies, but is cultivated extensively along the Senegal and Upper Niger rivers. In Gold Coast it is usually planted in June, alone or with yams, and harvested in December. In Niger it is planted in February and harvested in October. These tubers are eaten for food and also used as a remedy for dysentery.

White Potatoes. These are grown very little in the British colonies but somewhat more in the French colonies inland. There is confusion in the statistical tables between these and fura-fura potatoes, and the following table contains the figures for both:

Colony	*Acreage*	*Lbs. per acre*	*Crop in tons*
French Sudan	800	(3,000)	1,200
Niger		8,000–16,000	
Ivory Coast	1,240	(7,600)	4,700
Gold Coast		3,360	
Nigeria		1,200–7,200	
Cameroon	375		300,000

Ginger is grown in small plots in several of the coastal colonies, partly for local consumption but mainly for export. The chief crop is in Sierra Leone, with smaller amounts in Nigeria and Gold Coast. Over 2,700 tons were exported from Sierra Leone in 1938, and about 500 tons are annually exported from Nigeria. Considerable but smaller amounts are exported by the Ashanti tribes of the Gold Coast.

ONIONS: Albassa. Several species of onion including the Common Onion (*A. cepa*) of which there are many varieties, African Onion (*A. angolense*), and others are widely cultivated all over West Africa in small gardens around the huts, where other vegetables are grown and fertilized with ashes and house waste. They are also widely grown in irrigated lands along the rivers. Almost no figures are available as to the size of the crop, which is consumed locally; but in 1937, 600 tons were harvested in Niger.

TIGER NUTS: Earth Almonds. This is a native of southern Europe and North and West Africa, and the small nut-like tubers are gathered from wild plants. They have a sweetish taste and are eaten raw or roasted as a relish. Where the plants grow thickly, as in Yola, on the Benue River in northeast Nigeria, the yield of tubers may be as high as 7,000 pounds per acre.

CARROTS. These are grown to a small extent in irrigated gardens in Niger and in Gabon.

HERBACEOUS FRUITS AND VEGETABLES

A wide variety of herbaceous fruits and vegetables are grown throughout West Africa in small plots adjoining houses or in the villages. These are eaten by the families which grow them; and, although the total volume is very large, no record is kept of them and no statistics are available as to exact amounts, with a few exceptions. Among the more important are pineapple, chillies, and cayenne pepper, small amounts of which are exported from certain colonies; two species of black pepper; okra; tomatoes; eggplant; various cucurbits, including watermelon, squash, muskmelon, pumpkin, and cucumbers; and several salad greens such as sorrel, dock leaves, spiny amaranth, purslane, and basella.

OKRA: Gumbo, Bania. Okra is grown in almost every village in West Africa and forms an important food element everywhere. It is eaten as a vegetable and used for thickening soup. It is planted both as a field crop and near the houses in the villages. It is often cultivated as a field crop with yams, in April or May, and the fruit is picked two months later. In Ivory Coast, 8,800 tons were gathered in 1937, and it is probable that similar quantities were grown in the other colonies of West Africa. In Gold Coast, an experimental field in Ashanti yielded 778 pounds per acre.

EGGPLANT. This is an important secondary crop, especially in the Rain Forest belt, but is grown extensively by the Ashanti in northern Gold Coast and everywhere in West Africa where there is enough water.

TOMATOES. Tomatoes are among the most universally cultivated of the domestic vegetables. A large number of varieties are found in the native markets. They are grown throughout the Forest belt, and everywhere else where there is irrigation. Large crops are harvested in the southern Gold Coast, and along rivers in northern Nigeria and Niger. In Gabon they are grown both by natives and whites, and in Liberia they are cultivated extensively near Kassata. No figures are available for the yield or total harvest.

PUMPKINS. Pumpkins are grown everywhere in West Africa, interplanted or along fences. As with other cucurbits, the seeds are often roasted and eaten as a relish and are more esteemed than the flesh.

WATERMELONS. Watermelons are also raised everywhere. In Nigeria and Gold Coast (where they are planted in June and ripen in October to November) the seeds, called "neri," are more esteemed than the pulp, which is often discarded. In Niger and other places near the Desert, they are used as a source of potable water.

SQUASH, CUCUMBERS, and MUSKMELONS are also grown in many gardens in several of the colonies, as are several other melons and gourds. All are either interplanted with root crops or are grown along fences.

RED PEPPERS or Cayenne Peppers and CHILLIES or Green Peppers are universally raised as relishes. In descriptions of crops, they are generally considered together as "red pepper." They are grown in Nigeria both near houses and as field plants, where the yield (probably chillies) is 4,000 to 8,000 pounds per acre. In Ashanti chillies are extensively grown as a field crop, often under the shade of the locust bean, and when planted in May are ripe in November. In Liberia, small amounts are raised near Monrovia. In Sierra Leone, chillies are a minor export crop, almost 100 tons having been exported in 1938.

BENIN PEPPER: Black Pepper, White Pepper. *Piper guineense*, an indigenous species, is found wild throughout the wetter parts of West Africa and is also cultivated extensively. *Piper nigrum* is grown in the Gold Coast. Benin pepper is much used as a flavor but is not exported.

GUINEA PEPPER. This native pepper is found wild and is also raised widely. In Nigeria it is grown in small plots near houses. In Liberia it grows along the St. Paul River and near Fetoma. In Ivory Coast large quantities are gathered, 300 tons being exported in 1937.

Pineapples. Pineapples have been introduced in recent years into both the British and French coastal colonies, and in many have become a minor export crop. They are grown from the Rain Forest to the High Grass belt, and are eaten generally by the native population. From southern Nigeria, 3½ tons were sent to northern provinces in 1938. They are cultivated everywhere in Ashanti and are sold in quantity in the native markets of Gabon. Exports are chiefly from the French colonies, particularly Guinea, where more are grown than in all the others. The following amounts were exported in 1930:

French Guinea	40 tons
Ivory Coast	500 lbs.
Dahomey	500 lbs.

FRUIT TREES

Many fruit trees are grown or are native in West Africa, but most of them are found only as scattered specimens in the villages or forests; and, although their fruit may form a more or less important element in the diet, the total crop is not sufficient to make it worth while to determine the actual amount of fruit collected. Fruit trees of this kind, a few of which are likely to be found in or near any village, include the mango, pawpaw, guava, tamarind, custard apple, avocado, breadfruit, jujube, and monkey plum. Some figs are grown in Sierra Leone, and date palms are raised in the northern part of French Equatorial Africa. Vanilla is cultivated to a small extent in Nigeria and Gold Coast and in French Equatorial Africa.

The only important fruit trees of West Africa are the banana and plantain, but attempts are being made to introduce the cultivation of citrus fruits. These have been most successful with limes in Gold Coast and oranges in Senegal.

Bananas and Plantains. Bananas and plantains are grown throughout the coastal region of West Africa. They are usually planted together, and in compiling figures of acreages and crops, no distinction is usually made between them. In general, plantains are preferred by the natives as a regular food and bananas are largely gathered for export. Both are sold in quantity in the native markets of Gabon. In Sudan and northern Senegal only plantains are grown. In Ivory Coast plantains are one of the chief foods, and they are the main food of some of the inhabitants of Ashanti in north Gold Coast. In 1937, 490 tons of plantains were exported from Nigeria. In Liberia the greatest production is near Monrovia.

Bananas form one of the chief exports of several colonies, and the export is in addition to large amounts consumed locally.

The following table shows such figures as are available:

Colony	Acreage	Tons harvested	Tons exported
Senegal		400	
French Guinea		46,800	
Ivory Coast	120,000	360,300	
Gold Coast	2,400	1,000	
Nigeria			62,000
Cameroon		24,000	16,800

ORANGES. Oranges have been only recently introduced, and the climate has not been found very suitable for sweet oranges except in Senegal. The bitter orange has done better in southern Nigeria, and 349 tons of a small green variety, not suited to European trade, were sent to the northern provinces in 1938. In Dahomey cultivation has been more successful, and in Liberia they are grown near Monrovia and Bporo. In Senegal, 160 tons of satisfactory quality were harvested in 1938. In Gabon a great many are sold in the local markets.

LIMES. Limes have been extensively introduced in northern Nigeria and are grown in Liberia, near Monrovia and Bporo. The greatest success has been attained in Gold Coast where one plot of 2,000 acres produced 16,000 tons of fruit. Owing to poor storage facilities, greater success has been attained with export of lime juice than of the fruit. In 1937, 335,000 gallons of juice and 6,800 pounds of lime oil were exported in addition to 45 tons of fruit.

CITRON is grown in Niger and in the Forest belt of Nigeria.

GRAPEFRUIT is being introduced here and there along the coast, but little is yet raised.

OIL PLANTS

OIL PALM. The fruit of the oil palm is a large fleshy mass like a great pineapple, consisting of a large number of bright orange-colored drupes having a thin skin, thick pulp (pericarp), and a hard stone. Both the pulp and the stones, or nuts, contain oil.

Oil from the pericarp is an important native food and also used as a lubricant; the oil of the seed is used for cooking and for the preparation of margarine. The variety known as *macrosperma* has an especially thick pericarp with much oil. Certain varieties such as *tenera* have little pulp around the seeds. The oil in the pericarp becomes rancid if left in the shell and must be extracted when gathered. The kernels keep well.

The oil palm is not only the chief source of fat in the coastal region but also the greatest export crop. From many places it is the only export crop. It is a native tree and most of the oil is gathered from wild plants, although extensive plantations are also utilized. Both nuts and oil are exported in large quantities.

The table shows acreage, crop in tons, export of nuts and oil in tons:

Colony	*Acreage*	*Crop in tons*	*Export nuts in tons*	*Export oil in tons*
Senegal		3,500	3,000	77
Gambia			860	
French Guinea		19,300	19,000	450
Sierra Leone	9,000		84,500	1,200
Ivory Coast			11,000	7,000
Gold Coast			11,300	6,000
Dahomey		95,000	51,000	21,500
Togoland		21,500		3,300
Nigeria	8,700		386,000	132,000
Cameroon			44,300	10,300
French Equatorial			16,520	7,180

BENNISEED: Sesame, Tilseed. This is one of the most important sources of oil except in the Forest belt near the coast. It is an annual herb, the seeds of which contain large amounts of an edible oil. This oil is an important element of the diet throughout West Africa, and the seeds are also exported in large quantities. In Nigeria it is much grown by the Munshi tribe in Benue province, chiefly in small plots near houses. It is planted in July and harvested in February. In Niger it is widely grown in the loop of the Niger River.

The table shows data for some of the colonies:

Colony	*Acreage*	*Yield per acre*	*Crop in tons*	*Export in tons*
Senegal				50
French Guinea	22,200		900	
Ivory Coast			4,300	
Nigeria				18,000
Sierra Leone				400
Cameroon	2,700	100–200	1,700	
French Equatorial	22,980		3,100	

SHEA TREE: Butter Tree, Shea Nut. The shea tree is a native of West Africa, ranging from the Forest belt to near the Desert, and is also extensively planted.[1] The seeds contain about 50 per

[1] The shea tree is said to grow in a transverse belt from long. 16° W to 32° E, and from lat. 20° N to the Rain Forest.

cent of fat, which is obtained by softening the seed by fermentation and boiling. Both the nuts and the extracted oil are exported to Europe for margarine and soap. With benniseed, it is the chief source of fat except on the coast.

The following table shows available data in tons:

Colony	*Export seed in tons*	*Export oil in tons*
Senegal	348	1,482
French Guinea		11
Ivory Coast		2,100
Gold Coast	3,000 seed and oil	
Dahomey	1,300	200
Nigeria	6,700	
Cameroon	100	
French Sudan	2,900 seed and oil	

COCONUTS. The coconut is a minor source of fat in the diet and a minor export crop here and there along the entire Guinea coast. It is grown also along rivers inland, especially in Gold Coast. The crop is largely in white hands and most of it is exported as whole nuts, copra, and coconut oil.

Colony	*Crop in tons*	*Export nuts and copra in tons*	*Export oil (gallons)*
French West Africa	1,000		
Dahomey	300		
Togo		5,600	
Gold Coast		2,000	45,000?
Cameroon	2,600		
Nigeria		200	

SUGAR CANES

SUGAR CANE. Sugar cane is widely grown by the natives along the coast and the rivers of West Africa. The stems are chewed, but no sugar is made from it. In Nigeria it is often planted along the edges of rice fields. It is cut in November and yields about 15 tons per acre.

SWEET SORGHUM is cultivated by the native tribes, although it also grows wild, and the stalks are chewed for their sugar content. It is a sweetmeat rather than a food.

CAFFEINE PLANTS

The natives depend on kola as a source of caffeine, but in recent years cocoa and coffee have been introduced. Cocoa has become an important export crop, but not so much coffee is grown as in East Africa.

KOLA. The kola tree grows wild in the Rain Forest belt and is also much cultivated. The seeds, which are the part collected, contain about 1 per cent caffeine. The seeds are much used by the natives for preparing caffeine drinks; and there is a large internal trade in the seed, of which only a small part is exported.

Colony	*Export in tons*
French Guinea	117
Sierra Leone	2,250
Ivory Coast	2,500
Gold Coast	5,000
Togoland	240

COCOA: Cacao Tree. Cocoa is grown almost exclusively by the natives near the coast, but is hardly used by them—the entire crop being exported in normal times. As a wartime necessity various products of cocoa are now used locally. It is the most valuable money crop of West Africa, exports being exceeded in tonnage only by the products of the oil palm. More than half the world's supply now comes from West Africa. The fruit is collected in the late fall, when the amount of fat in the seeds is greatest, amounting to 55 per cent. The seeds contain theobromine, an alkaloid related chemically to caffeine. Cacao cultivation has been particularly encouraged in British West Africa as a native enterprise, while in Cameroon it is grown on large European plantations started by the Germans.

Colony	*Square miles*	*Crop in tons*
French Equatorial	60	1,000
Sierra Leone	200	6,000
Ivory Coast	1,300	53,000
Gold Coast	1,500	285,000
Dahomey	100	30
Togo	200	16,000
Nigeria	2,000	100,000
Cameroon	600	31,000
Spanish Guinea		14,200

COFFEE. Three species of *Coffea* are commonly grown in West Africa: two are native African species, one Arabian. They are Congo Coffee (*C. robusta*), Liberian Coffee (*C. liberica*), and Mocha Coffee (*C. arabica*). Almost all is consumed locally, but it is a minor export crop from several of the colonies and from Liberia. In Nigeria all species are grown extensively, but the crop is consumed by the native population or exported to other colonies. In French Guinea where some coffee is grown, local consumption is still larger than local production. In Ivory Coast over 17,000 acres were under cultivation in 1928; and in 1940, 13,000 tons were exported. In Gold Coast the crop is about 50 tons, and in Sierra Leone slightly less. From French Equatorial Africa, 2,600 tons were exported in 1940. Export is steadily increasing from a small start in 1934. Almost all is from Middle Congo.

Chapter II

EQUATORIAL AFRICA

The territories of the Belgian Congo and Angola differ in two respects from those of West Africa. Large areas have been opened to white proprietors as plantations, and they also include enormous undeveloped tracts of forest and savannah occupied by native tribes which have little contact with the white settlers. The entire tract is watered by the Congo River and its tributaries, and the white settlements have developed chiefly along the coast and on the navigable rivers. There are no accurate reports of agricultural conditions except in these areas. With the exception of the southern half of Angola, which is Park Savannah, the entire area is tropical Rain Forest. Over much of the area the tsetse fly does not occur and mixed farming is possible. The crops are similar to those of West Africa, but the southern and inland parts are cooler and crops characteristic of temperate regions are grown more frequently. Owing to the presence of cattle, manuring is possible, and the practice of allowing the land to revert to wilderness after a few years of cultivation is therefore less prevalent. Mine products and animal products, such as hides, are more important than agricultural crops as exports, and much less attention has been paid to developing agriculture than in West Africa. For this reason data on crops are much less complete, and only scattered observations by travelers and occasional figures on exports are available. Cereals, starchy roots, and legumes form the basis of the native diet, in addition to milk, meat, and animal foods. There are many large plantations under white ownership—especially along the Congo, as at Leopoldville, Stanleyville, and Equatorville—and on these a great variety of fruits and vegetables are grown, chiefly for local consumption. There is much more distinction between native and European plantings than in West Africa, and figures are often given for each separately. The figures on native plantings are probably much less accurate than those for other parts of Africa. It is loosely stated that the figures as given should be about doubled.

CEREAL CROPS

The cereal crops are much the same as in West Africa; but the millets, which are grown in dry regions, are planted less often.

Guineacorn and bulrush millet are less important, and maize and rice are grown more than farther north.

GUINEACORN is cultivated all along the Congo and its tributaries and in the savannahs in the Uele, Katanga, and Kasai regions of the Congo, occupying about 250,000 acres. It is raised in the northern part of Angola, but not much in the southern half.

BULRUSH MILLET is grown widely on the steppes of the Congo and Kasai, occupying about 80,000 acres, and by some of the tribes in Angola, but not by the Ovimbundu of southern Angola.

MAIZE is the chief cereal crop of the entire region. It is raised extensively along the Congo River and on the intervening steppes. Throughout Angola it is the chief cereal, and among the Ovimbundu tribe in the south, almost the only one.

Colony	*Native acreage*	*White acreage*	*Crop in tons*	*Export in tons*
Belgian Congo	700,000	5,000	34,000*	4,000
Angola	1,600,000		500,000	170,000

* This figure should be much increased as it is the crop from only Ruanda-Urundi, where the acreage was 174,000 acres.

RICE. This is a very important crop both in the Belgian Congo and in Angola. It is grown chiefly along the rivers, and most of the crop is consumed locally or sold to nearby tribes. In the Belgian Congo much is grown near Stanleyville and Leopoldville, and by Arabs in Aruwimi in the extreme northeast.

Colony	*Native acreage*	*White acreage*	*Crop in tons*	*Export in tons*
Belgian Congo	250,000	250	60,000*	1,200
Angola	18,000		8,200	2,300

* This should probably be about 100,000 tons as it represents the yield from only 150,000 acres.

TAMBA is raised in the savannahs of the eastern and southern parts of the Belgian Congo.

WHEAT. Wheat is grown very little by the natives in the Belgian Congo, but small amounts are cultivated by the whites for their own use. In Angola large crops are planted on the Benguela plateau. It is used by the whites, and its use by the natives is increasing. In 1938, 2,400 tons were exported. In 1933, the crop is said to have been over 34,000 tons, harvested from 25,000 acres.

LEGUMES

Legumes are probably grown less than in West Africa, but most of the kinds planted there are found to some extent in the West Coast Colonies.

Haricot Beans. These are cultivated widely in Belgian Congo, both along the rivers and in the savannahs, but are not an important crop. In Angola a crop of 25,100 tons was gathered in 1939 from 125,000 acres. Of this 5,678 tons were exported. In 1941, 11,241 tons were exported.

Lima Beans, Pigeon Peas, and Bambara Groundnuts are also grown to some extent in the savannahs in Belgian Congo, and Peas (*Pisum sativum*) on white plantations near Leopoldville and by the Ovimbundu in Angola.

Peanuts. In the Belgian Congo peanuts are an important crop in the savannahs and in Mayomba. In 1939, 60,500 tons were harvested from 250,000 acres, and over 6,000 tons were exported in 1936. In Angola peanuts are grown extensively everywhere. In 1939, 7,100 tons were harvested from 25,000 acres. In 1941, 254 tons were exported.

ROOT CROPS

Starchy roots, together with maize, constitute the chief diet of the natives of the West Coast Colonies. Yams, cassava, and sweet potatoes are the most important. Coco-yams are grown infrequently. Fra-Fra potatoes are an important secondary food, and white potatoes are raised in considerable amount on European plantations and by the natives for sale to the white residents.

Yams. In Belgian Congo, the Water Yam is grown everywhere. There are large plantings along the branches of the Congo River, in villages near Stephanieville, in Mayomba near Lake Tanganyika, and in the sandy parts of the northwest. In Angola there are enormous plantations in the south, both in Mossamedes and among the Ovimbundu. No figures of acreage or crop are available.

Cassava. In Belgian Congo it is the chief native food in the entire Congo River basin, and is the most important root crop in Mayomba and in the sandy northeast sector. In 1938, native plantings were 800,000 acres; European, 6,250. Export was 23,810 tons. In Angola it is the chief food crop, both in the forested part and among the Ovimbundu. In 1939, 301,590 tons were harvested.

Sweet Potatoes. In Belgian Congo, they are secondary only to cassava, very large crops being grown near Leopoldville, in Upoto, in Aruwimi, and in Mayomba. Native plantings in 1935 occupied 160,000 acres; European, 3,500 acres. The European crop was 9,800 tons; the native, probably over 200,000. Export in 1938 was 8,730 tons. In Angola it is an important crop in the Congo basin and also among the Ovimbundu, but no figures are available.

FRA-FRA POTATOES. In the Belgian Congo they are grown in the savannahs. In 1938 native plantings occupied 1,900 acres.

WHITE POTATOES. In the Belgian Congo they are grown very little except in Ruanda-Urundi, north of Lake Tanganyika. In 1939, 62,500 acres yielded 215,000 tons. In 1937 there was an export of 1,477 tons gathered from 650 acres, on European plantations. In Angola they are raised in small quantity by the Ovimbundu for sale to whites, and on white plantations on the plateau.

GINGER. Large crops are raised in the upper Lupweshi district near Lake Tanganyika.

ONIONS. African onions are grown widely everywhere. Large crops are gathered along all the Congo River branches and by Arabs in Aruwimi. By the Ovimbundu in southern Angola, they are cultivated chiefly for sale to whites.

CARROTS and RADISHES are grown in the Belgian Congo, in the white plantations near Leopoldville, and in Angola by the Ovimbundu. BEETS are raised by the Ovimbundu in Angola.

HERBACEOUS FRUITS AND VEGETABLES

A very wide variety of these is grown on the white plantations and by the natives for sale to the whites as well as for their own use. The most widely grown appear to be tomatoes, eggplant, and chillies. On the white plantations in Belgian Congo sufficient amounts of such vegetables and fruits as cabbage, celery, and chicory are grown for their own needs. In Angola the Ovimbundu sell to the whites large amounts of cabbage and strawberries, which are also grown on white plantations. Many cucurbits are raised everywhere. Various greens are eaten including spinach, amaranth, and basella.

EGGPLANT. A small fruited form is raised everywhere in Belgian Congo.

TOMATOES. These are grown extensively everywhere, notably in Upoto and Mayomba, and by the Arabs in Aruwimi in the Belgian Congo and by the Ovimbundu in Angola.

CHILLIES. These are cultivated all over the colonies, especially in the Belgian Congo near Leopoldville, Stephanieville, and in Upoto and Mayomba. They are much used by the Ovimbundu in Angola. The small cayenne pepper is grown infrequently.

PINEAPPLES are raised in large amount in forest and savannah in the Belgian Congo, and both in the forest regions and by the Ovimbundu in Angola.

FRUIT TREES

A greater variety of fruit trees is grown than in West Africa. The tropical fruit trees, including bananas, plantains, mangoes, breadfruits, and custard apples are cultivated everywhere; and in addition, in Belgian Congo, we find pawpaws, citrus trees, guava, avocado, and others. In Angola the southern part is cool enough for fruits of temperate climates; and in addition to those mentioned, we find lemons, tangerines, guava, pears, apples, peaches, mulberries, and figs. Grapes grow wild both in Belgian Congo and Angola, and are sold in the local markets.

BANANAS AND PLANTAINS. These fruits are the chief foods in some of the forest areas of the Belgian Congo. They are grown all along the Congo River, in Upoto, and in Mayomba. They do not grow north of the Makowa River in Aruwimi. In 1937 there were 400,000 acres of native plantings and 4,300 of white, indicating a probable crop of over 500,000 tons, of which 12,146 tons were exported. In Angola it is one of many fruits eaten by the natives.

ORANGES. In Belgian Congo oranges are grown chiefly by whites near Leopoldville, by Arabs in Aruwimi and in Mayomba. In Angola they constitute an important crop both in the forest regions and in the area of the Ovimbundu.

MANGO. These are raised everywhere. They are cultivated especially on white plantations near the Congo in Belgian Congo and are grown by the Ovimbundu in Angola.

PAWPAW. The pawpaw grows wild in all the forest districts and is raised on European plantations on the Congo River. It is also cultivated by the Ovimbundu in Angola.

OIL AND SUGAR PLANTS

Oil plants are less numerous than in West Africa. The oil palm extends down the coast to the middle of Angola, but there are fewer trees than on the Guinea coast. Sesame is cultivated to some extent, and coconuts are grown in a small area of northern Angola.

OIL PALM. In the Belgian Congo, oil palms grow along most of the Congo and Aruwimi rivers and on the shore of Lake Tanganyika. Plantations established by Europeans cover 64,000 acres, in addition to almost 400,000 acres of native trees. From the European plantings, 92,000 tons of kernels and 60,000 tons of oil were exported in 1937. Most of the crop gathered by natives is consumed locally. In Angola, oil palms grow near the coast south to Benguela, on the Cuanza River up to 150 miles inland, and on the Kasai River near Bemba. Most of the oil is consumed locally, but from 1936 to 1940 the export trade averaged 6,300 tons of nuts and 3,000 tons of oil.

SESAME. Sesame is an important crop everywhere in both colonies, most of the crop being consumed locally. In the Belgian Congo, native plantings covered 27,500 acres in 1936, and 300 tons of seed were exported. From 1934 to 1938, the crop is said to have decreased from 6,100 to 1,000 tons. From Angola there are conflicting reports regarding sesame. Although from 1934 to 1938 the crop is said to have fallen from 1,600 to 700 tons, export increased from 110 to 970 tons; and this is reported to reflect an increase in the total crop. It seems evident that the reports of crop are much too low in recent years.

COCONUT. Coconuts are grown infrequently in Belgian Congo. In Angola, there are small stands along the Cuanza River and on the coast north of this. In 1939, 6,560 tons of copra were exported, but only 4,500 in 1941.

SUGAR CANE. This is a much more important crop than in West Africa; and under white influence much sugar is produced, both in Belgian Congo and Angola. In Belgian Congo it is raised chiefly on white plantations in swamps near Stephanieville. In Angola it grows in marshy ground near rivers and at Catumbela.

The table shows acreage, crop, sugar production, and export in tons.

Colony	*Acreage*	*Sugar*	*Export cane in tons*	*Export sugar in tons*
Belgian Congo	8,000	3,350	9,936	14,200
Angola	35,000			38,200

CAFFEINE PLANTS

The crops of cocoa and kola are less than those of West Africa, but that of coffee is greater.

KOLA. This is found wild in the forests of the Belgian Congo, especially along the Congo River below Stanley Falls. The crop of wild seed is about 1,200 tons from 15,000 acres. In 1936, 500 tons were exported. Recently white plantations of 8,000 acres yielded a crop of 9,936 tons, of which 8,476 tons were exported in 1936.

COCOA. Little is grown in Angola or Belgian Congo. In the latter there are about 12,500 acres of European plantations along the Congo River, giving an annual crop of 1,100 to 1,500 tons. In Angola, cocoa cultivation is limited to the Cabinda enclave. Here 3,400 acres are under European ownership and even more under native. In 1939, 324 tons were exported.

COFFEE. Coffee is an important crop in the Belgian Congo. The largest plantations are near Leopoldville and Equatorville, totaling 7,500 acres European and 5,600 native. In 1937, 18,000 tons were exported and in 1942, 32,000 tons were harvested. In Angola, *C. robusta* grows wild along the coast and *C. arabica* is cultivated inland, for example at Cassinga, Encage, and Novo-Redondo. It is the chief export crop, 17,600 tons being exported in 1939.

Chapter III

NORTH EAST AFRICA

North East Africa comprises Anglo-Egyptian Sudan, Ethiopia, and Somaliland (French, British, and Italian Somaliland). The whole of Somaliland is too dry for satisfactory agriculture, and plant crops are limited to a strip of land 80 miles long and 2 to 15 miles wide in Hargeisa and Borana, in the western half of British Somaliland. Crops are limited to maize, millet, beans, and a little wheat and barley. In French and Italian Somaliland, most vegetable food is imported. The natives have large flocks of goats and sheep and herds of cattle and camels. Milk, cheese, and fish form the bulk of the diet. In Ethiopia similar conditions prevail. Large quantities of hides, butter, soap, and other animal products are exported. Durra, a form of guineacorn, is the principal crop; barley and wheat are grown to a less extent. Along the rivers a great variety of crops is planted, chillies and potatoes being the most widely cultivated. Coffee is the only export crop, but much of it is of poor quality, known as Abyssinian coffee. The beans are small and are often picked off the ground and are mixed with dirt. A smaller amount of Mocha coffee is grown more carefully and exported. Laws forbid the export of other vegetable crops. The Anglo-Egyptian Sudan includes a wide variety of types of land. The Nile flows through the eastern half and the White and Blue Nile Rivers join a little above the center of the country. Along the Nile the land is irrigated and very fertile, and the tongue of land between the rivers, known as "el Gezira," is the source of much export of durra, dukhn (a form of bulrush millet), cotton and cottonseed, sesame, peanuts, and dates. The northwestern part of the country is largely desert, where dates and millets are the chief crops. The southern part is typical of central Africa, having mountains (the Nuba Mountains), rivers, and swamps. The rainfall is 30 to 43 inches. Durra is the chief crop but there is much maize, tolabun (*Eleusine corocana*) and dukhn. Shea trees are common.

CEREAL CROPS

GUINEACORN, called durra or dhura, is the chief crop over most of the region but is largely imported in Somaliland. In Ethiopia it is the chief food crop. In Anglo-Egyptian Sudan very large amounts are grown in some provinces, and it is even exported.

The crop varies from 200,000 to 500,000 tons annually. The export from Gezira is over 33,000 tons every year.

Bulrush Millet, called dukhn or dhokn, is second only to durra. It also is grown over most of the region. It is cultivated in the Hargesia and Borana districts of Somaliland but is little grown in Ethiopia. In Anglo-Egyptian Sudan it is grown less than durra, except in the mountainous districts of Kordofan. In Dar Nuba, it is replaced by maize. In 1930 the total crop in Sudan was 118,000 tons. From Gezira 150 tons are exported annually.

Maize. It is one of the chief crops of the fertile strip in Somaliland, and in Ethiopia is grown to a small extent near Harrar. In Anglo-Egyptian Sudan it is grown chiefly in Dar Nuba and along the Nile in Halfa and Dongola. From 1937 to 1940 the average crop was 10,300 tons from 28,000 acres.

Wheat. In Somaliland a little is grown on the fertile strip, and in Ethiopia fairly large amounts are raised in various parts. In Anglo-Egyptian Sudan it is cultivated along the Nile in Halfa and Dongola and in White Nile Province. In 1942, 18,000 tons were harvested from 20,000 acres. In Abyssinia it is grown extensively in a few scattered localities.

Barley. In Somaliland a little is grown, but it is an important crop in Ethiopia. In Anglo-Egyptian Sudan it is raised along the Nile in Halfa and Dongola, and 3,000 tons were harvested in 1930.

LEGUMES

Several kinds of beans and peas are grown in small amounts in Anglo-Egyptian Sudan. Haricot Beans are raised along the Nile, and in 1939, 8,100 tons were harvested. Lablab Beans are grown extensively in southern Bahr el Ghazal and in irrigated land along the Nile. Mungo Beans are cultivated in Bahr el Ghazal and Lentils in Sennar between the White and Blue Nile. White beans grow in Kordofan, the crop exceeds 200 tons. The Dinka tribes grow Voandezia.

Peanuts are grown in many places in Anglo-Egyptian Sudan; and in 1939, 48,000 acres were planted, of which 5,000 were in Kordofan and 1,000 in White Nile. There were 8,600 tons exported in 1939 in addition to 10,000 from Gezira alone. In addition to the exports much is consumed locally.

ROOT CROPS

Root crops are grown very little in this part of Africa and nowhere form an important part of the food. Yams, coco-yams, and cassava are grown to some extent in the southern part of Bahr el Ghazal, and white and sweet potatoes chiefly on government and military posts in the same province. White potatoes are also raised to some extent throughout Ethiopia. Onions are cultivated in the provinces of White Nile, Kordofan, Darfur, and Gezira in Anglo-Egyptian Sudan, but not in Bahr el Ghazal.

HERBACEOUS VEGETABLES AND FRUITS

In Anglo-Egyptian Sudan the favorite vegetables are cucurbits of various kinds. Cucumbers, pumpkins, watermelons, and certain native species including a kind of muskmelon and a large gourd are grown widely in eastern and central Kordofan, along the Nile around Halfa and Dongola, in Darfur, in White Nile, and in Bahr el Ghazal.

OKRA is grown in the wetter parts of Kordofan, in Maridi, in Bahr el Ghazal, and in irrigated land along the Nile.

TOMATOES and other vegetables are grown in all provinces, particularly in the gardens of military and government officials.

PINEAPPLES are raised extensively in Maridi in the south.

CHILLIES are widely cultivated in Ethiopia.

FRUIT TREES

Little fruit is grown except in certain parts of the Anglo-Egyptian Sudan. Between the White and Blue Niles, where irrigated gardens are feasible, figs, lemons, pomegranates, and bananas are not uncommon in the summer months. Bananas are native in the southern part of Bahr el Ghazal, and pawpaws grow in Maridi.

DATES are the only important fruit crop of the country. They constitute the chief article of diet of the native tribes in the western desert regions of Halfa and Dongola and the northern part of Berber district. They are also grown along the Nile above Atbar. In 1924 export amounted to over 4,000 tons.

OIL AND SUGAR PLANTS

SESAME is one of the chief crops in all the watered parts of Anglo-Egyptian Sudan. In Kordofan there are 28,000 acres; in White Nile 27,000 acres; and in Nuba, Fung, Kassale, and Bahr el Ghazal upwards of 70,000 acres. Benniseed is an important item of export from the Nile Valley in Halfa and Dongola. From Gezira the average export is over 9,000 tons. The total crop was given as 37,687 tons in 1930, and as 43,000 tons in 1938. Small amounts also come from Darfur.

SHEA TREE. Many shea trees are found in Bahr el Ghazal between the Tonj and Naaru rivers. It extends from about lat. 12° to 20° N. and as far east as long. 32° E. (see footnote, p. 27).

SUGAR CANE. Sugar cane is not an important crop, but in Ethiopia 500 acres were planted on the Baro River near Gambela.

CAFFEINE PLANTS

COFFEE is extensively grown in Ethiopia and is the only food exported. There are two species: Mocha coffee which is carefully cultivated and yields good beans, and Abyssinian coffee which is indigenous and has a small bean of poor flavor. This is often picked from the ground and is mixed with dirt. It is consumed locally or exported to nearby provinces. In 1938, 19,800 tons of Mocha coffee were exported.

Of crops other than foods, cotton is by far the most important with gum acacia a poor second. Both are of more value than any of the food crop exports.

COTTON is grown in many parts of Anglo-Egyptian Sudan and exports reach about 36,000 tons a year, of which about one-fourth is from Gezira. Export of cotton seed is about 18,000 tons annually.

GUM ACACIA comes chiefly from Kordofan, and about 3,500 tons are exported annually.

Chapter IV

EAST AFRICA

East Africa—comprising Kenya, Uganda, Tanganyika, Nyasaland, and Northern Rhodesia—differs markedly in its topography from West Africa. Instead of the vast stretches of forest, savannah, and desert, with a few great rivers, the country is one of numerous lakes, high mountains, valleys, and numerous smaller rivers. Agriculture is much more diversified, and it is not possible to define certain great zones characterized by definite agricultural possibilities. The Indian Ocean coasts of Kenya and Tanganyika bear certain resemblances to the western coast, and in the forests along them many of the same food trees are found, such as oil palms, coconuts, and mangoes. Bananas and plantains grow near the lakes more than along the coast. The native-grown cereals are much the same as in West Africa, but wheat appears to gain in popularity. Among root crops, cassava and sweet potatoes are cultivated more than the yams. White potatoes are grown widely. Peas and beans of many kinds flourish, and peanuts are among the chief crops. Citrus and other fruit trees are not important food or export crops. Benniseed is the chief source of fat in much of the area. Sugar cane is grown extensively along the coast and the rivers, and much sugar is produced. Cocoa is being introduced in Tanganyika. Coffee is grown much more than on the west coast, and tea is cultivated in the mountains. Kola is rare or absent. In Northern Rhodesia grazing largely replaces other agriculture.

Large parts of all the colonies are cultivated by white settlers, partly in great plantations, partly on small holdings. The crops in such are chiefly those to which Europeans are accustomed, such as rice and wheat among the cereals, and white potatoes in the root crops. White settlers grow at least half of the sugar cane and coffee and almost all of the tea.

Cotton is grown extensively, and in Northern Rhodesia tobacco is the chief export crop.

In Kenya, some of the native tribes hold cattle as an evidence of wealth and gather herds which are larger than the forage of the region can supply, thus aggravating the already serious problem of soil erosion. Goats are particularly destructive, as they eat young trees and thus prevent reforestation. The Kamba, for example, maintain 190,000 head of cattle with grazing land sufficient for only 60,000 head.

The table below shows approximate acreage of European and native cultivation in recent years. The figures for native plantings are incomplete and should perhaps be doubled.

Colony	*European*	*Native*
Kenya	600,000	1,800,000
Uganda	30,000	4,000,000
Tanganyika	2,500,000	4,000,000
Nyasaland	60,000	60,000
Northern Rhodesia	70,000	Not known

CEREAL CROPS

GUINEACORN. Guineacorn, although extensively grown in all the colonies, is less important as a food than in West Africa. It is the chief cereal of Northern Rhodesia, except on the Great Plateau, but in the other colonies is less grown than bulrush millet or maize. In Uganda the area planted is about 260,000 acres, and in Kenya about the same. There are no estimates for crops or for the other colonies.

BULRUSH MILLET. This is the chief cereal crop in Kenya, Uganda, and Tanganyika, the southern part of Nyasaland, and among the Wemba tribes of Northern Rhodesia. In Tanganyika, 1,700,000 acres yield an annual crop of about 500,000 tons, of which 15,000 tons were exported in 1935. In Uganda it flourishes especially in the south, near Lake Usoga. In the whole colony plantings reach 980,000 acres. In Kenya, about 250,000 acres are grown.

TAMBA. Tamba is an important element of the diet all over Uganda, especially in the eastern and northern parts where 880,000 acres are planted. It grows also in the southern part of Nyasaland and in the northeast part of Northern Rhodesia, where it is the chief cereal crop on the Great Plateau. It is cultivated less in Kenya and Tanganyika, the crop from the latter being 8,200 tons in 1937, chiefly from Western Province.

MAIZE. Maize is an important crop everywhere, although grown less than the millets. In Kenya, Nyasaland, and Tanganyika, it is a minor export crop. In Uganda it is cultivated especially near Lake Victoria; in Nyasaland, in the Shire highlands and near Zomba; and in Northern Rhodesia, on European plantations in the Kafue Flats, the Luangwa valley, and near Fort Jameson.

Colony	*Native acreage*	*White acreage*	*Crop in tons*	*Export in tons*
Kenya	386,310	245,807	208,200	45,000
Uganda	130,944	118		
Tanganyika	538,400	1,000	297,000	3,580
Nyasaland	4,700		(2,200)	5,400
Northern Rhodesia		42,757	29,000	

Rice. Rice is grown near the lakes and along the rivers in all the colonies and in Northern Rhodesia on the Great Plateau. In Tanganyika it grows near Lake Victoria and along the Rufiji River; in Nyasaland, by the Shire River and the shore of Lake Nyasa. Almost all is cultivated by natives. The following table shows such statistics as are available.

Colony	*Acreage*	*Crop*	*Export in tons*
Uganda	21,831		
Tanganyika	102,390		82,000
Nyasaland			1,400

Wheat. Wheat is an important crop among the Europeans, especially in higher altitudes. It is becoming increasingly important in the European settlements of Kenya and Northern Rhodesia particularly along the railways. In Kenya, plantings in 1938 comprised 57,000 acres European and 914 native, yielding a crop of about 25,000 tons, of which 18,000 tons were exported; in 1942 production rose to 131,000 tons. In Uganda, 1,353 acres were reported in 1937. Tanganyika had 13,320 acres of native plantings in 1929, in addition to European. In Northern Rhodesia approximately 1,000 tons are harvested every year from over 2,000 acres of white plantations. In Nyasaland where the small group of European settlers concentrate on tea cultivation less than 400 acres of wheat are planted.

Barley. Barley is one of the chief European crops in the highlands of Tanganyika and Kenya. In the latter territory over 2,500 acres are under cultivation, yielding a crop of 1,400 tons. Only small areas of barley are cultivated in the other colonies.

The following table shows maximum or approximate figures since 1933 (in tons):

		Guinea-corn	*Bulrush millet*	*Maize*	*Rice*	*Wheat*
Tanganyika	acreage		1,700,000	540,000	102,000	57,000
	crop		500,000	30,000	(6,500)	24,900
	export		15,000	3,600	82,000	18,000
Uganda	acreage	264,000	980,000	131,000	21,800	1,350
	crop					
	export					
Kenya	acreage	500,000		630,000		13,300
	crop			200,000		
	export			45,000		
Nyasaland	acreage			4,500		400
	crop			(2,200)	(200)	80
	export			5,400	1,400	
Northern Rhodesia	acreage			42,000		2,000
	crop			29,000		1,000
	export					

LEGUMES

Peas and beans of many kinds are grown in East Africa, including haricot beans, white and navy beans, lima beans, peas, pigeon peas, cowpeas, soy beans, and various native beans. In most agricultural references these are not separated, and statistics are given for all legumes exclusive of peanuts. Although haricots and pigeon peas are specifically mentioned as being important crops in Nyasaland and haricots and cowpeas in Northern Rhodesia, it cannot be assumed that they are any less important elsewhere. Soy beans are grown in Nyasaland chiefly as a cover crop, therefore only 200 tons were exported in 1938.

Beans and Peas. Very few estimates regarding production are available, but the importance of the crops is indicated by the acreage planted, which has been as follows in recent years: Kenya, 305,414; Uganda, 329,140; Tanganyika, 550,000; Nyasaland, white plantings, 1,860. In Northern Rhodesia, cowpeas occupied 3,845 acres (European) and haricot beans, 685 acres.

Peanuts. Peanuts are grown widely in Uganda, especially in the eastern and northern parts. In Kenya, Kisumu Province supplies most of the export trade. In Tanganyika they are cultivated most extensively in the southern Lindi and western lake districts. They are less important in Nyasaland and Northern Rhodesia.

Colony	*Acreage*	*Crop in tons*	*Export in tons*
Kenya	13,000		3,900
Uganda	150,000		1,460
Tanganyika	60,000	39,000	22,250
Nyasaland	5,000	(700)	1,100
Northern Rhodesia (Europeans only)	1,430	(1,400)	

ROOT CROPS

Root crops are cultivated less extensively than in most other parts of Africa. The only ones of importance are cassava and sweet potatoes. Yams are an important secondary crop in the banana region of Tanganyika, near the lake, and in parts of Nyasaland. Coco-yams are grown in the same part of Tanganyika and in Emba in western Kenya.

Cassava. Both species of cassava are grown in the western part of Kenya, in Emba, and in the northern part of Tanganyika, near Lake Victoria. Cassava is one of the chief foods in most of Uganda, Nyasaland, and Northern Rhodesia. In Uganda, 360,000 acres were reported in 1937. In Tanganyika a crop of 670,000 tons is gathered from about 200,000 acres.

SWEET POTATOES. Sweet potatoes are extensively grown, both by natives and whites, in all the colonies. In Tanganyika, about 120,000 acres are planted every year; and in Uganda, over 500,000 acres. No figures are available for the other colonies.

WHITE POTATOES. These are grown to a moderate extent, both by natives and Europeans, in all the colonies. In Kenya, the natives harvest about 8,000 tons each year, chiefly from near Lake Victoria. In Tanganyika, over 3,000 acres are planted each year, yielding about 2,900 tons. In Uganda, over 6,000 acres are planted. In Northern Rhodesia, 200 tons were harvested from 125 acres in 1939, which was not the entire crop. Small amounts, up to 45 tons, are exported from Nyasaland.

HERBACEOUS VEGETABLES AND FRUITS

These are grown very little in East Africa, with the exception of chillies and various cucurbits. In 1940 there were reported 2,000 acres of "vegetables" in Kenya.

CHILLIES are grown especially in Uganda and Nyasaland. In Uganda they are raised chiefly in the banana plantations and in Busoga. In 1937, 657 acres were planted and 700 tons exported.

In Nyasaland and Kenya, chillies are grown both by whites and natives. Most of the crop is consumed locally. In 1933, 239 tons were exported from Nyasaland and 140 from Kenya, but the export has been much less in recent years.

CUCURBITS are favorite crops in Tanganyika, Nyasaland, and Northern Rhodesia. Pumpkins are planted on 7,000 acres in Tanganyika, and various melons occupied 740 acres of European plantings in Northern Rhodesia. Pumpkins and muskmelon are cultivated extensively on the Shire River in Nyasaland.

FRUIT TREES

Several fruit trees, especially bananas and plantains, mangoes and pawpaws, grow wild in various parts of East Africa, and the cultivation of citrus trees is being introduced.

BANANAS AND PLANTAINS. These are grown widely in Kenya, Uganda, and Tanganyika. In Kenya there are over 50,000 acres of native plantations, chiefly near the coast, yielding a crop of 8,000,000 bunches. In Uganda, plantain is the chief food of a large part of the population in the south near Lake Victoria. There are about 1,000,000 acres. In Tanganyika they grow near Lakes Victoria and Tanganyika, where plantains are one of the chief native foods occupying 250,000 acres. From Tanganyika over 5,000 tons are exported annually.

MANGOES abound in various places, especially on the coast of Tanganyika near Dar es Salaam. They are nowhere an important element of the diet.

PAWPAWS grow wild here and there and are cultivated extensively in Tanganyika and by the Wasengi in Northern Rhodesia.

ORANGES are grown on white farms along the railway in Northern Rhodesia and Kenya; from the latter 4,500,000 oranges were exported in 1942. GRAPEFRUIT are cultivated in Nyasaland both for local consumption and export. In 1938, 19 tons were exported.

OIL AND SUGAR PLANTS

OIL PALM. The oil palm, which dominates the exports of West Africa, is found only in a few places along the coast of Tanganyika. In 1929 there were 1,000 acres of oil palm plantations. No oil or nuts are exported.

SESAME. Benniseed and sesame oil are important native crops over most of Kenya, Uganda, and Tanganyika. Sesame is less grown in Nyasaland and Northern Rhodesia.

The table shows figures for a good year.

Colony	*Acreage*	*Crop in tons*	*Export seed in tons*	*Export oil in tons*
Kenya	48,000	3,500	3,300	440
Uganda	214,000	228,000	1,620	
Tanganyika	52,000		4,800	100

COCONUTS. Coconuts are widely cultivated by natives and whites on the coasts of Kenya and Tanganyika. In Kenya, 37,000 acres yielded a crop of 43,000 tons, from which about 450 tons of copra and 100 tons of oil were exported. In Tanganyika, the export of 7,500 tons of copra represents part of the yield from 45,000 acres.

SUGAR CANE. Sugar cane is grown in all the colonies, along the coast and near rivers. Much sugar is produced. In Kenya it is raised chiefly in Nyanza and Kikuyu and less on the coast. In 1938, 15,500 acres of European and 21,870 of native each yielded almost 160,000 tons of sugar, of which about half was exported. In Uganda, in the same year, there were 15,410 acres of Asiatic and 2,246 of native plantings. Most of the sugar produced was consumed locally, only 55 tons being exported. In Tanganyika, 24,500 acres yielded 7,000 tons of sugar, of which 4,000 were exported. In Nyasaland, small plantings are made along the Shire River and in Northern Rhodesia. A little is grown by the Wasenga tribe in Northern Rhodesia.

CAFFEINE PLANTS

Coffee is an important crop everywhere. Tea is grown on the southern slopes of the mountains. Cacao was introduced near the coast and on the lakes but has not been very successful. Little if any is exported.

COFFEE. In Kenya, *C. arabica* is grown chiefly. There are about 117,000 acres of plantations, with a yearly export of 30,000 tons. In Uganda, in 1939, there were 18,000 acres of white and 16,000 acres of native plantations. Yearly export is steadily increasing and reached 17,700 tons in 1940. Tanganyika has about 28,000 acres of white and 78,000 acres of native plantings, chiefly *C. robusta.* Exports in recent years have varied from 18,500 tons in 1935 to 11,000 tons in 1937. In Nyasaland, coffee is grown chiefly in Mlange and the Shire highlands. The crop is very uncertain because of variable weather. In 1939 there were 1,331 acres of white and about 200 of native plantings. Exports have varied from 400 tons in 1934 to 18 tons in 1936. The average of six years was 155 tons. In Northern Rhodesia cultivation is beginning, 162 acres were planted in 1939.

TEA. Kenya now produces about 5,000 tons yearly, most of which is exported. In Uganda only 342 acres are productive, and the crop is consumed locally. In Tanganyika it is grown chiefly in Rungwe, Mufindi, and on the southern slope of the highlands. The annual crop is about 260 tons, of which more than half is exported. In Nyasaland, tea is grown on the southern slope of Mlange Mountain and other highlands, occupying about 15,000 acres. Export in recent years has varied from a high of 6,000 tons in 1942 to 580 tons in 1937.

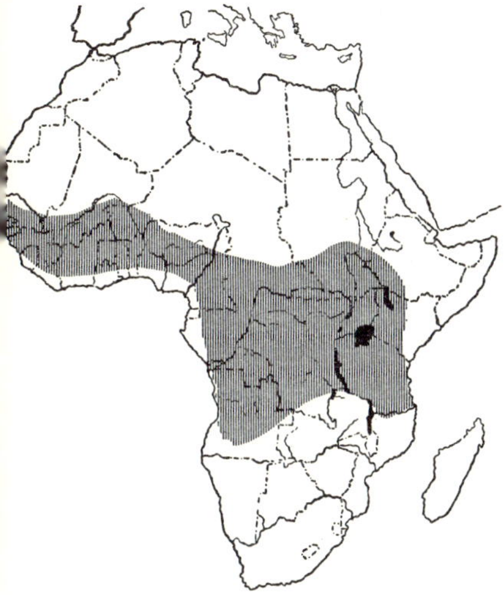

FIG. 2
BENNISEED (*Sesamum orientale*)

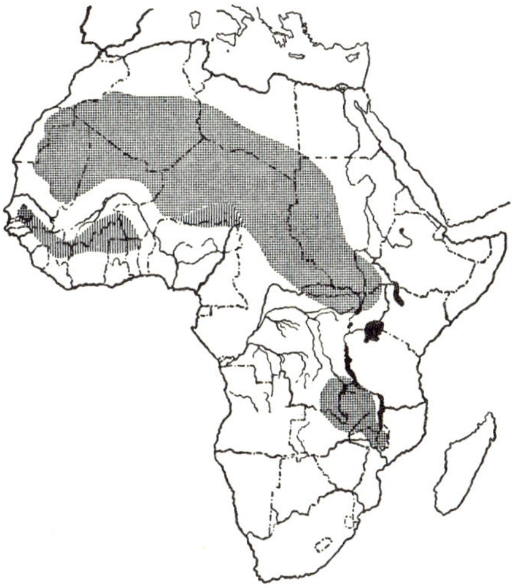

FIG. 3
LESSER MILLETS
TAMBA (*Eleusine corocana*)
FONIO (*Paspalum longiflorum*)
MAIWA (*Pennisetum spicatum*)
ACHA (*Digitaria exilis*)
IBURN (*Digitaria iburua*)

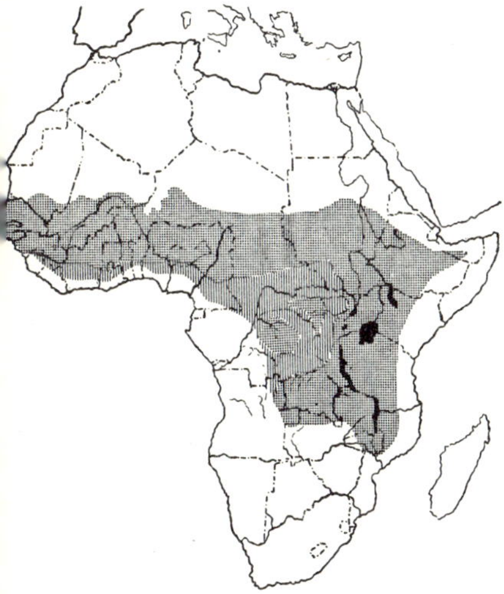

FIG. 4
BULRUSH MILLET
(*Pennisetum typhoideum*)

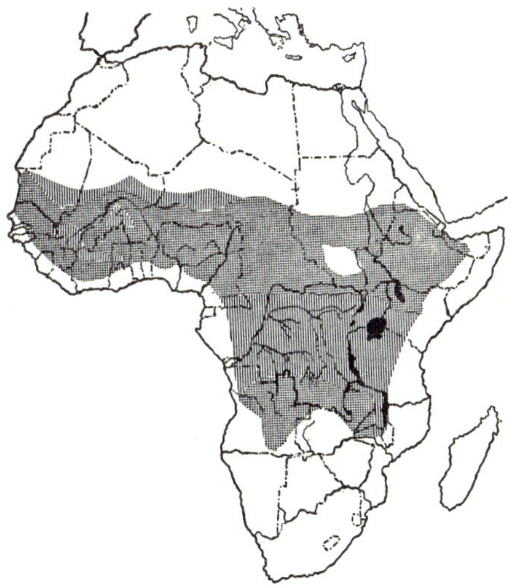

FIG. 5
GUINEACORN (*Sorghum vulgare*)

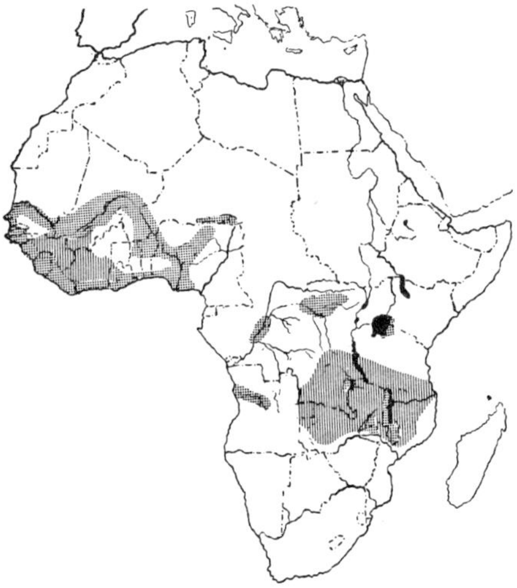

FIG. 6
RICE (*Oryza sativa*)

FIG. 7
WHEAT (*Triticum aestivum*)

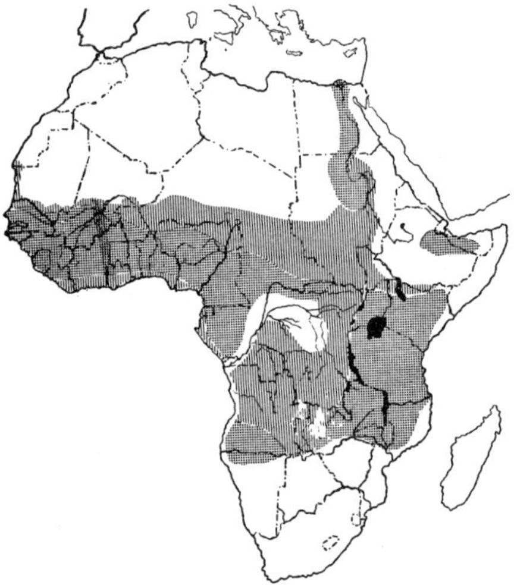

FIG. 8
MAIZE (*Zea mays*)

FIG. 9
PEANUT (*Arachis hypogea*)

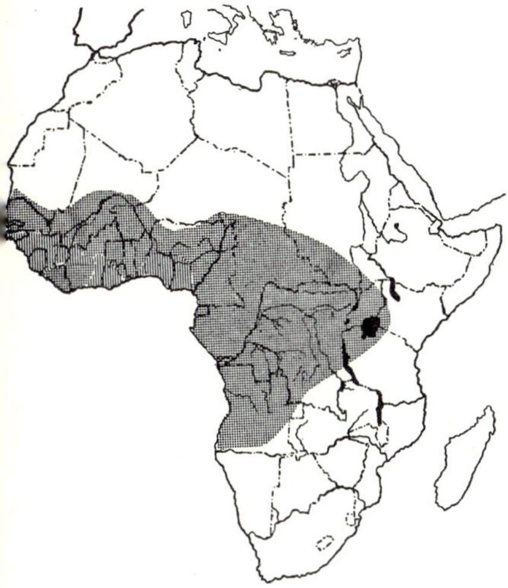

FIG. 10
CASSAVA (*Manihot*)

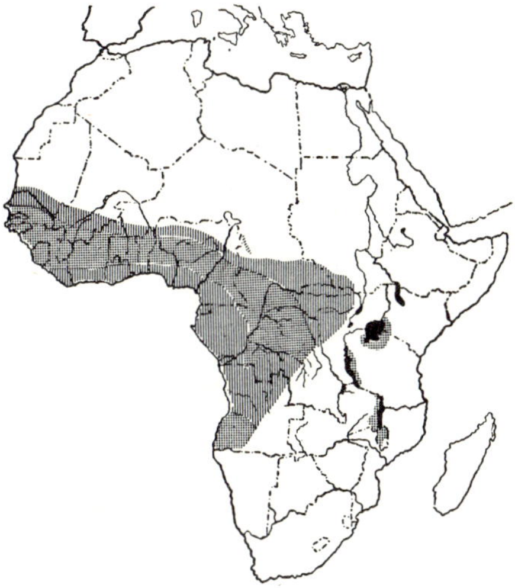

FIG. 11
YAMS (*Dioscorea*)
COCO-YAMS (*Colocasia esculenta*)

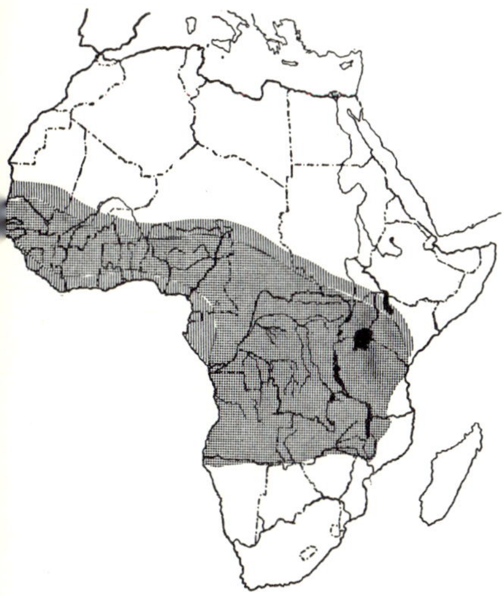

FIG. 12
SWEET POTATO (*Ipomoea batatas*)

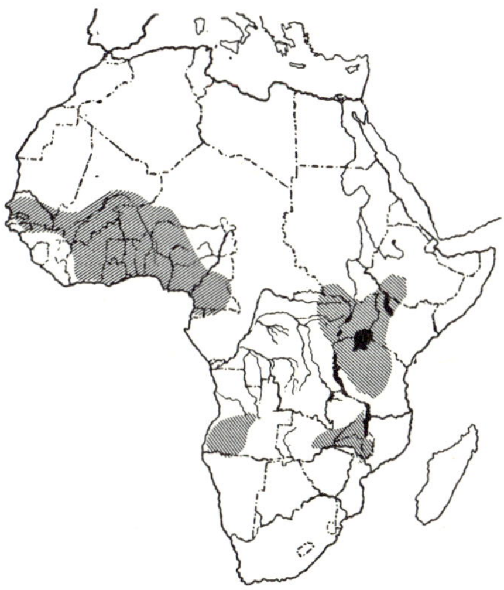

FIG. 13
WHITE POTATOES
(*Solanum tuberosum*)
FRA-FRA POTATOES
(*Coleus rotundifolius*)
These are sometimes confused in descriptions of products.

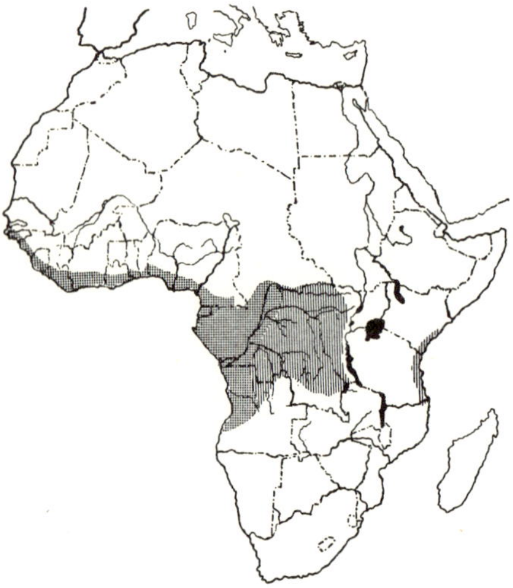

Fig. 14

Oil Palm (*Elaeis guineensis*)

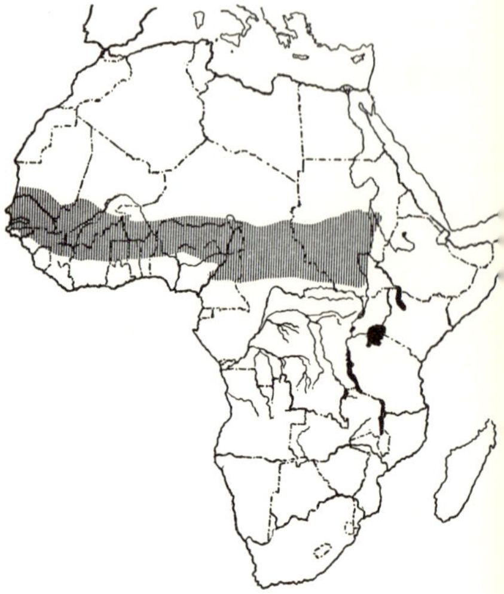

Fig. 15

Shea Tree (*Butyrospermum parkii*)

Fig. 16

Coconut (*Cocos nucifera*)

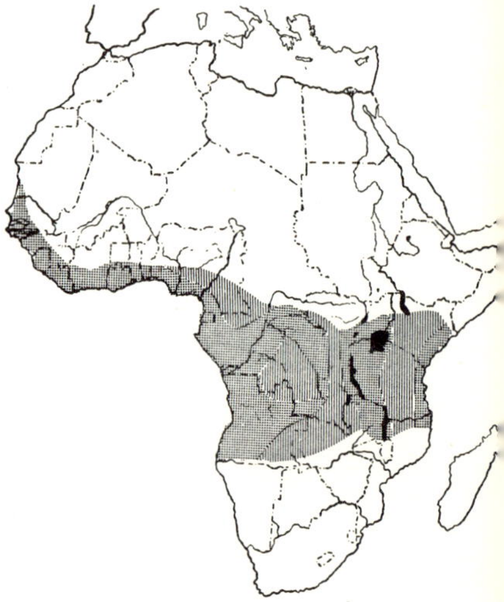

Fig. 17

Banana (*Musa sapientum*)
Plantain (*Musa paradisiaca*)

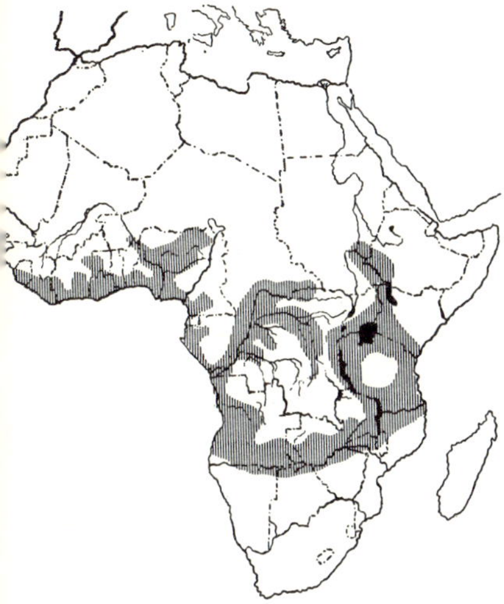

Fig. 18

Sugar Cane
(*Saccharum officinarum*)

Fig. 19

Cocoa (*Theobroma cacao*)

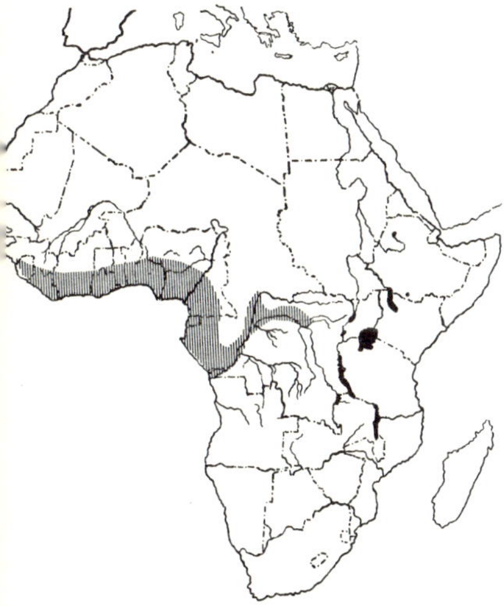

Fig. 20

Kola (*Cola acuminata*)

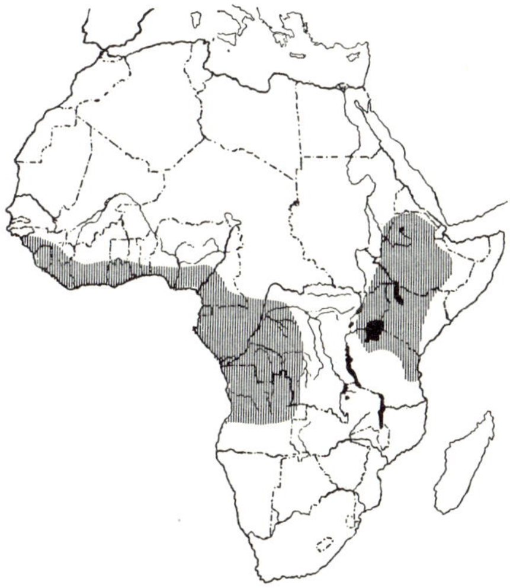

Fig. 21

Coffee (*Coffea*)

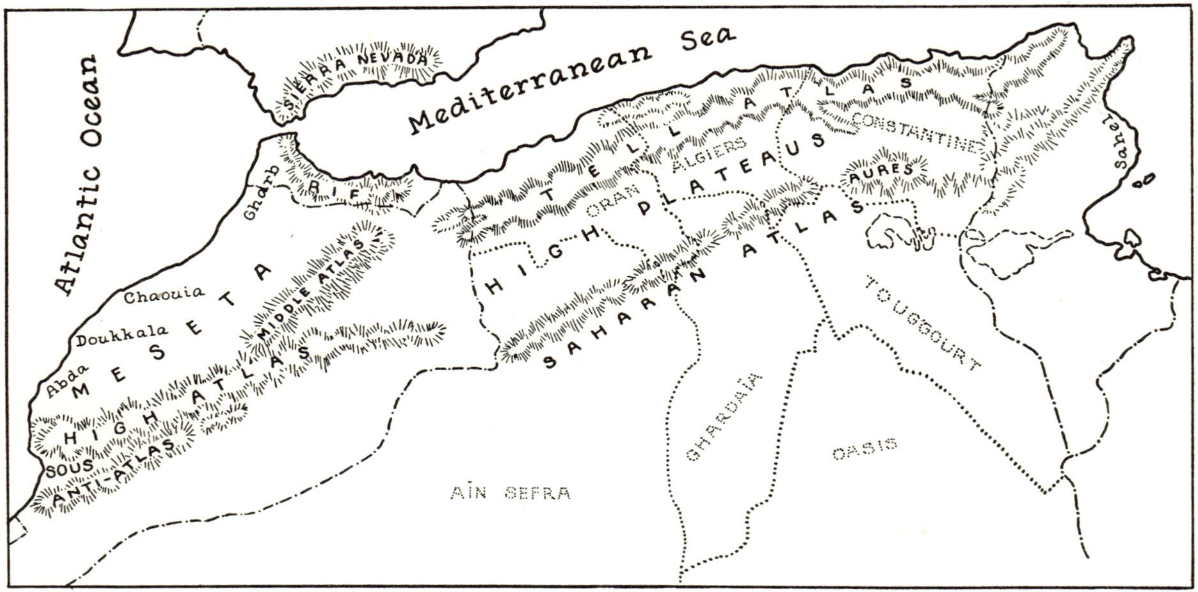

Fig. 22

North Africa

Showing the trends of the main Atlas ranges and political and geographical subdivisions of Morocco, Algeria, and Tunisia mentioned in the text. Each range here indicated as a ridge is in reality a complex mountain system. The political divisions of Algeria are indicated by dotted letters: the Departments of Oran, Algiers, and Constantine include most of the arable land; Aïn Sefra, Ghardaïa, Touggourt, and the Territory of the Oasis are largely desert and compose for administrative purposes the Southern Territories.

NORTH AFRICA: INTRODUCTION

By

Carroll E. Wood, Jr.

North Africa is essentially more a part of the Mediterranean region than it is a part of Africa. In its flora, fauna, climate, and physiographic structure, it is very clearly related to the European continent and so constitutes a unique portion of Africa.

The climate is of the Mediterranean type so characteristic of the Mediterranean Basin and the west coasts of most other continents: the southwest coast of Africa, southwestern Australia, Chile, and southern California. This peculiar type of climate is characterized by cool, wet winters and hot, dry, nearly or quite rainless summers as in North Africa where practically all of the 10 to 40 inches of precipitation takes place during the winter six months when the prevailing winds are from the west and northwest.

The flora, too, is of a special Mediterranean type, widespread in the lands around the basin of the inland sea and adapted to withstand the conditions to which it is subjected. Trees and shrubs with persistent, usually small, thick leaves predominate; bulbous or thorny plants are common types. Leaves of plants are also frequently whitened, waxy, or hairy, or with some other modification which serves to prevent loss of moisture during the summer period. The cork oak, characteristic of the western Mediterranean and found in northern Morocco and the coastal ranges of Algeria and northern Tunisia, serves well to illustrate the Mediterranean type with its thickened, small, evergreen leaves and heavy, waterproof bark as protection against excessive evaporation. Throughout the forested areas, which are now largely confined to mountain slopes, oaks and cone-bearing evergreens predominate, and in the underbrush the arbutus tree, myrtle, broom, species of dwarf oak and other shrubby plants also illustrate the general climatic conditions. It is these undershrubs which, after the forest is cut away, become the tangled scrub, known as *maquis*, now thought of as so characteristic of Mediterranean shores. Only in the higher mountains of Morocco and Algeria are conditions of moisture more equable the year around, due partly to the winter snows which melt slowly through the summer. It is in those regions that the magnificent Atlas cedar (*Cedrus libanotica*, var. *atlantica*), species of maple, and other types more characteristic of moist, cool climates are found. Elsewhere the Mediterranean vegetation is predominant.

In general, with the exception of the high mountain ranges rainfall decreases inland, from north to south. As rainfall decreases, the Mediterranean type of vegetation gradually passes into a grassland (steppe) formation in which alfa grass (*Stipa tenacissima*), an important natural source of fiber, is particularly conspicuous. Across the high plateaus of eastern Morocco, Algeria, and Tunisia, great areas of this grass occur in regions of relatively low rainfall. The steppe, in its turn, gives way to the desert, where vegetation is even more severely limited and modified by the lack of rainfall and by greater temperature extremes.

North Africa is essentially an agricultural region. It has been deficient in fuel supplies and water resources so that little industrial development has taken place, and agriculture in one form or another is the major occupation. The Maghreb, as the region is called in Arabic, was the ancient granary of Rome, and at the beginning of the present war was well on its way to becoming an important exporter of agricultural products.

But the agriculture of the region is, like the flora, limited by the climate. In this warm, dry climate, almost all temperate and subtropical crops are grown or could be grown, but the lack of rainfall during the summer is a limiting factor. The intensively cultivated lands are concentrated in the comparatively well-watered coastal belt, while in the interior the land is given over to cereal culture or grazing. Of the entire area under cultivation, however, only 10 per cent (including the rather extensive oases of the Sahara) is under irrigation, so that the remaining 90 per cent depends on the winter wet period which decreases in length inland and southward. With the exception of tree crops, therefore, half of the year when sunshine is greatest is lost to agriculture.

Under these conditions cereals and tree crops are especially suited to the Mediterranean area, for cereals can mature by early summer and trees with their extensive root systems are able to tap water reserves, thus better resisting the dry period. Hard and soft wheat and barley are the main cereals, and the olive, almond, and fig may be cited as conspicuous examples of tree crops important in the region. (Although citrus fruits and dates are important crops, they are grown only in irrigated areas.)

It should be noted that in addition to being peculiarly distributed in time rainfall is often very irregular from year to year, resulting in erratic yields, frequently with serious consequences. For example, in 1917–18, a good agricultural year with an average of more than 20 inches of rainfall on the Tell regions of Algeria and corresponding amounts inland, the harvest of cereals reached 2,185,000 tons. However, in 1919–20, a bad year, with less than 12 inches on the Tell, the harvest amounted to only 951,000 tons. This is not an isolated case, for irregular harvests are the rule, and the effect of climatic conditions on olive production is so great that it is almost impossible to speak of an "average" yield. The serious effect of drought was well illustrated by the dry years 1935–37, which caused serious disturbances among the peoples of the oases along the edge of the Moroccan Sahara, in the High Atlas, and elsewhere.

Aridity is therefore the main limitation to agriculture, constituting a permanent problem for North Africa. The irregularity of crop yields could be compensated for by adequate storage facilities which are in many regions lacking, and irrigation could compensate for the low and variable rainfall. Some steps have already been taken in this direction, and the Moroccan government had outlined a comprehensive, long-term plan, which was to have begun in 1940, for the full utilization of the water resources of the country.

The agricultural picture in North Africa is further complicated by the presence of both European and native (Arab and Berber) farmers whose interests frequently conflict. Native methods of agriculture are primitive, and the local populations are inclined to resist changes in the traditional methods. However, since the coming of the French, the agricultural situation has been alleviated somewhat and modern agricultural practices are gradually being adopted. Agricultural experiment stations have been set up in all three territories, and new techniques and crop varieties have been developed. Many European farmers in the region make use of the most advanced mechanical equipment. Fuel shortages since 1940 have greatly hampered agricultural production, however.

With the exception of tea, coffee, and sugar, most of the food consumed in North Africa is ordinarily produced in that region and there is usually a surplus of cereals, wine, olive oil, citrus, and other fruits and vegetables. This surplus disappeared very shortly after 1940 when the Axis began draining the Maghreb of all available food supplies, and it has not been seen since. However, quantities of essential materials are now reaching the region through the agency of Lend-Lease and the Army.

Under normal conditions, sugar is the main deficiency; although small irrigated patches of cane are grown in the Sous Valley of southwestern Morocco and around Marrakech, Moroccan production of sugar is practically nil and all sugar is imported. Morocco formerly produced sugar, and it is likely that it could become at least partially self-sufficient again by the replacement with cane of the scattered banana groves of Sous and elsewhere along the edge of the desert (precariously established at best because of the danger of frost) and by the extension of sugar culture in other regions.

In addition, that part of western Morocco known as the Gharb and the area around Rabat have been shown to be suitable for the growth of sugar beets in rotation with wheat. The sugar beet has the advantage that it grows during the winter months and would not have to be irrigated.

Tea and coffee cannot be grown in North Africa, for the Mediterranean climate is generally unsuited to tropical crops. Moroccans consume great quantities of green tea, and Algerians of coffee, and these two items must continue to be imported. With these exceptions North Africa could be and should be virtually self-supporting and should under ordinary conditions be able to supply surpluses of many crops to the Allied Nations.

The account of the various important food plants in North Africa which follows is not, nor is it intended to be, comprehensive. Limited by available space, it is intended instead to provide a somewhat general picture of what may be expected of the region in more normal times. Only the three French territories of most interest at the present time are considered; Spanish Morocco and Libya are hardly touched upon.

The material included here is taken from sources available in this country and presents mostly a pre-Vichy France picture. Statistics included are those available and are usually given for several years to show the fluctuations to which North African harvests are subject. Most statistics are taken either directly or indirectly from French sources and should, perhaps, with the exception of exports, be used as a guide and not as the absolute truth. The necessity for statistics in a bureaucracy has apparently brought forth material on all crops in all French colonial territories, but it should again be noted (see Introduction, Part I) that it is quite impossible to get complete estimates of area and production, especially when areas are small and scattered. Since most of the Moroccan figures are the result of surveys for taxation purposes, it is likely that these are underestimates. The same is true of Algeria and Tunisia.

In addition to the food crops discussed here, there are a number of plants grown for industrial purposes which are not considered in detail. Tobacco is the most important, especially in the coastal areas of Algeria and in a few of the Tunisian oases. Flax is grown for the seed in the plains of western Morocco, but production varies according to conditions on the world market. Cotton may also be mentioned, but it does well only under irrigation and has not been an important crop in North Africa since production was resumed in the South following our Civil War. Cultivated fiber plants such as hemp and sisal are of little importance, but "vegetable hair" obtained from the leaves of the dwarf palm which grows wild in dry plains, particularly in the Department of Oran, was prior to the present war an important export—up to 50,000 tons from Algeria. Alfa grass, which is abundant on many of the steppes of Morocco, Algeria, and Tunisia, was before the war exported in large quantities, mostly to England for paper manufacture.

Classification of Lands in North Africa *

	Algeria	*Tunisia*	*Morocco*
ARABLE LANDS	*5,840,000 hectares*[a]	*2,935,000*	*7,077,000*
Cereals	3,207,000	1,650,000	3,600,000
Legumes (peas, beans)	93,000	38,000	165,000
Industrial crops (Tobacco, flax, cotton, etc.)	26,000	830	40,000
Land under fallow	2,740,000	1,100,000	3,000,000
Meadow and feed crops	42,000	35,000	36,000
TREE AND BUSH CROPS	*3,630,000*	*400,000*	*193,000*
Olives	82,000	17,360[b]	7,657[b]
Vineyards	400,000	42,500	23,000
Citrus	10,000	300[b]	1,340[b]
Figs	62,000		7,000[b]
Date palms	57,480 (6,800)[b]	2,675[b]	1,000[b]
Others (almond, apricot, carob, apple, pear, cherry, chestnut, etc.)	10,700	2,500[b]	5,000[b]
PERMANENT MEADOWS AND PASTURES CUT FOR HAY	*35,000*	*100,000*	
FORESTS AND BRUSH	*3,630,000*	*1,016,000*	*2,600,000*
UNCULTIVATED PRODUCTIVE AND UNPRODUCTIVE LANDS	*210,350,000*	*8,050,000*	*29,993,000*
(Unproductive land)		(3,500,000)	(19,743,000)
(Grazing lands)	(20,000,000)		(7,958,000)
	220,486,000	12,451,000	39,863,000

* Figures from the International Institute of Agriculture for 1937, the most recent estimate. Modified.

[a] 1 hectare = approximately 2½ acres (2.47 acres).

[b] Thousands of trees—other figures in hectares.

Crops Grown in North Africa

Cereals, page 64
- Hard wheat
- Soft wheat
- Barley
- Oats
- Maize
- Sorghum
- Rye

Legumes, page 71
- Broad beans
- Kidney beans
- Soybeans
- Carobs
- Chickpeas
- Fenugreek
- Garden peas
- Lentils

Root Crops, page 72
- White potatoes

Vineyards, page 73
- Wine grapes

Oil Plants, page 78
- Olives
- Argan

Tree Crops, page 82
- Almonds
- Apricots
- Bananas
- Citrus fruits
  - Citrons
  - Lemons
  - Limes
  - Mandarins
  - Grapefruit
  - Oranges
  - Tangerines
  - Kumquats
- Cherries
- Chestnuts
- Dates

Tree Crops, page 82
- Figs
- Medlars
- Pears
- Peaches
- Plums
- Pomegranates
- Quinces
- Walnuts

Market-garden (Truck) Crops, page 89
- Globe Artichokes
- Asparagus
- Cabbage
- Carrots
- Green beans
- Celery
- Fennel
- Garden peas
- Melons
- New potatoes
- Onions
- Squash
- Tomatoes
- Table Grapes
- Turnips
- Watermelons

Herbs and Spices, page 91

Miscellaneous non-food crops not considered here
- Castor beans
- Cotton
- Hemp
- Henna
- Flax (seed)
- Perfume plants (Rose geranium, lavender, citronella, bitter orange)
- Sisal
- Tobacco

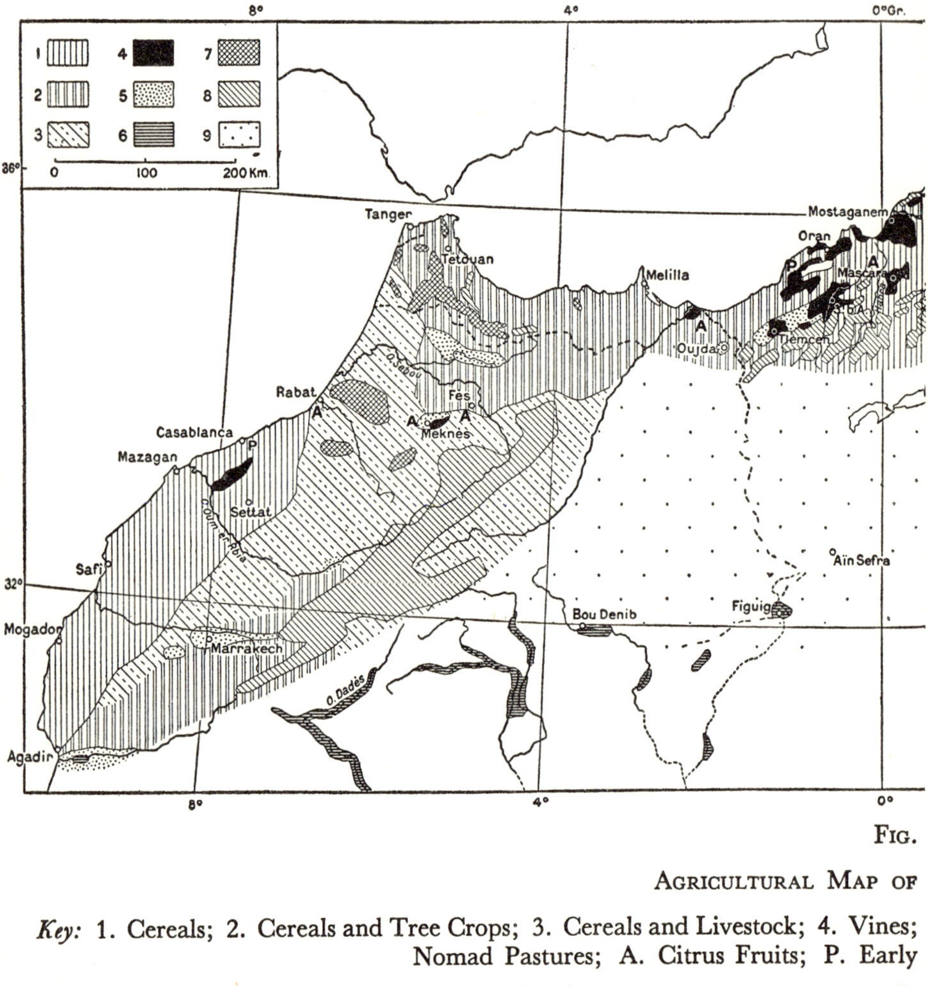

FIG.

AGRICULTURAL MAP OF

Key: 1. Cereals; 2. Cereals and Tree Crops; 3. Cereals and Livestock; 4. Vines; Nomad Pastures; A. Citrus Fruits; P. Early

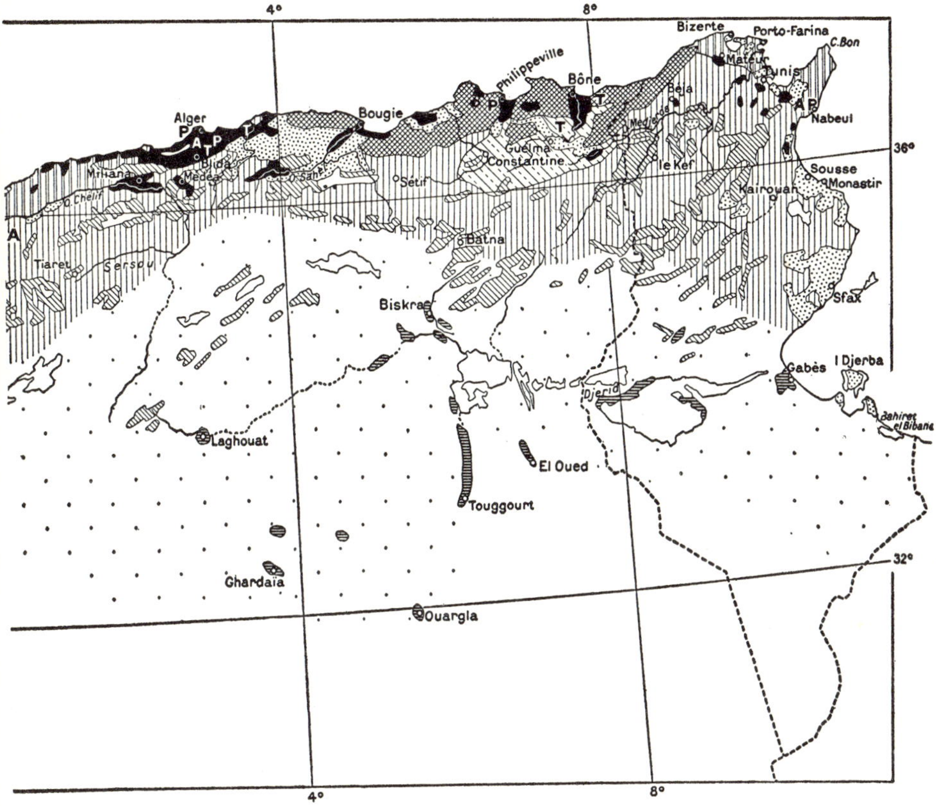

23

French North Africa

5. Olives; 6. Palms; 7. Cork Oak Forests; 8. Other Forests; 9. Steppes and Vegetables; T. Tobacco. Scale 1: 7,700,000.

Chapter V

FRENCH NORTH AFRICA

CEREAL CROPS

Cereals are cultivated nearly everywhere in North Africa and from the standpoint of both area and production are more important than any other type of crop. In western Morocco they occupy some 3,000,000 hectares; in Algeria a like acreage, mainly in the Tell region and on the high plateaus; in Tunisia 1,000,000 hectares. Although the area planted in cereals varies comparatively little from year to year, irregularities in rainfall produce great fluctuations in yield, frequently with serious results.

Of the several cereals grown, hard and soft wheat and barley are most important. Hard wheat is grown mainly by the natives, but the production of soft wheat under European impetus is increasing. Barley is planted by the natives almost everywhere and is particularly important in regions where the rainfall is too low for successful wheat production. Oats, rye, maize, and sorghum are less extensively grown and constitute a much smaller part of the total.

Since the coming of the French in the middle of the last century and the introduction of soft wheat with its higher yields, the cultivation of cereals has been put on a much sounder basis. New varieties of wheat and barley particularly adapted to North African climatic conditions have been developed, and in addition special agricultural techniques, among which "dry-farming" is conspicuous, have been brought into use. Under this method of farming, the land which would ordinarily lie fallow for a year between crops of wheat is prepared at the beginning of the fallow year. The weeds are destroyed, the soil is enriched in nitrogen, and the penetration of the winter rains is facilitated by the breaking up of the soil so that water is trapped there. In this way the wheat planted the next year benefits by the rains of two years, becoming, perhaps, in a sense a biennial plant. Dry-farming has been practised mostly by Europeans with their mechanized farm equipment; the natives, on the other hand, simply let their lands lie bare during the off year. Steel plows, however, and more modern farm tools are gradually replacing the primitive agricultural instruments of the region, and in addition the more successful European practices are being adopted with a consequent increase in crop yields.

WHEAT. Both hard and soft wheats are widely cultivated in North Africa wherever rainfall is sufficient. In most regions, wheat production is limited by the isohyet of 15 inches of rainfall; where rainfall is less than this, barley is grown. Small quantities of wheat are produced in the desert oases during the winter.

The great wheat-producing areas of Morocco are the western plateaus and plains (the meseta), including the Chaouïa, the Doukkala, the Sebou Basin and the Gharb, and the regions around Fez and Meknès—in general the clayey soils formerly covered by a scrub of wild olive, mastic (*Pistachia lentiscus*), and dwarf palm (*Chamaerops humilis*). Annual production ranges between 500,000 and 1,000,000 tons (630,000 in 1938).

In Algeria, the rich plains of the Tell region are preferred by Europeans for the culture of wheat, but the high plateaus south of the Tell Atlas ranges are the real wheat regions—the plains of the Tell now being given over mostly to more intensive cultures such as citrus, vine, tobacco, and truck crops. Although yields, especially inland and on native farms, are frequently low, the aggregate production amounts to 800,000 to 1,100,000 tons yearly, an average of between .5 and .7 tons per hectare.

In Tunisia, wheat is extensively grown in the plateau and northern regions but is replaced southward by barley as rainfall diminishes. Tunisian production ranges between 200,000 and 460,000 tons.

Hard wheat, with its relatively great resistance to drought and high temperatures, is well adapted to North African climatic conditions, and it has been suggested that the region is one of the centers of origin of wheat of this type with its large, very hard grains, rich in gluten. This was the sole type of wheat cultivated in North Africa before the coming of the French and is still widespread among the native populations which it furnishes with a valuable food supply, being utilized in making the ubiquitous cous-cous and edible paste products, on the order of macaroni and spaghetti, such as dweeda.

Soft wheat, with its higher yields per acre and greater market value, is being more and more widely grown in the Maghreb. Europeans grow this type almost exclusively; and each year more native farmers, quick to see the advantages of this grain, are planting it instead of the traditional hard wheat.

In Tunisia, in 1920, some 52,000 hectares of soft wheat, producing 34,800 tons, were grown; in 1938, 160,000 hectares, yielding 180,000 tons of soft wheat, were planted—a rather astounding increase in acreage and production. Hard wheat, on the other hand, occupied

WHEAT PRODUCTION IN NORTH AFRICA *

	1934		1936		1938	
	Area	*Produc-tion*	*Area*	*Produc-tion*	*Area*	*Produc-tion*
	hectares	*tons*	*hectares*	*tons*	*hectares*	*tons*
ALGERIA						
Hard wheat			130,800	5,076	1,230,000	627,000
Soft wheat			42,600	3,027	430,000	323,000
Total	1,646,000	1,184,100	173,400	8,103	1,660,000	950,000
MOROCCO						
Hard wheat	854,000	757,000	975,662	231,000	837,960	365,000
Soft wheat	321,274	320,000	316,787	102,000	375,582	265,200
Total	1,221,000	1,077,000	1,292,449	333,000	1,214,000	630,700
TUNISIA						
Hard wheat	668,000	200,000	360,000	120,000	514,000	200,400
Soft wheat	120,000	175,000	134,000	100,000	161,000	180,000
Total	788,000	375,000	494,000	220,000	675,000	380,400

* From *International Year Book of Agricultural Statistics.* Modified.

480,000 hectares in 1920 (107,500 tons) and 514,000 hectares (200,400 tons) in 1938.

The same increase in soft wheat production has taken place in Morocco, where in 1920 only 15,000 hectares of soft wheat, with a yield of 10,400 tons, were planted. In 1938, however, 375,000 hectares produced 265,200 tons. Similar figures emphasize the increase in production of soft wheat in Algeria, but here again, hard wheat is still the most widely grown type.

The effect of fluctuations in rainfall from one year to the next on wheat and cereal production should be emphasized. A recent example may be noted in Morocco, where 321,200 hectares planted in soft wheat in 1934 produced 320,000 tons of grain. A similar acreage (303,000 hectares) in 1936 gave only 90,000 tons.

Production in North Africa is usually sufficient, however. Total production of wheat in 1939–40 was 2,700,000 tons and in 1940–41, 1,700,000 tons. Nineteen forty-two was generally a dry year with relatively poor harvests throughout North Africa. It is understood that low rainfall in Morocco and warfare in Tunisia have caused the harvest to fall below normal again in 1943.

PRODUCTION OF SOFT WHEAT IN MOROCCO *

(Area in hectares. Production in metric tons.)

	Europeans		*Natives*		*Total*	
	Area	*Production*	*Area*	*Production*	*Area*	*Production*
1920	9,720	6,828	5,500	3,581	15,220	10,409
1925	31,418	30,344	57,684	49,145	89,103	79,490
1928	65,374	52,299	154,294	59,492	219,667	111,791
1931	81,416	87,151	129,182	127,848	210,598	215,000
1934	124,710	144,339	197,064	175,655	321,274	320,000
1936	125,000	50,000	178,000	40,000	303,000	90,000

* From *L'Encyclopédie coloniale et maritime*, Series IV, fasc. 6, 1937–38.

A surplus of wheat, ranging between 100,000 and 600,000 tons, is usually available and was formerly exported according to a quota system to France. The table below shows exports to 1938, the last year available.

WHEAT EXPORTS FROM NORTH AFRICA
(In metric tons)

	Algeria	*Morocco*	*Tunisia*
1928–32	203,660	127,540	169,620
1933	233,520	217,140	76,700
1934	315,100	164,600	58,490
1935	265,170	177,450	153,940
1936	261,430	82,070	29,530
1937	161,270	47,160	68,360
1938	96,760	113,870	82,200

BARLEY. Barley is of great antiquity and was cultivated even before wheat; some strains of barley may have originated in North Africa. The grain of barley differs from that of wheat in that it is closely invested by the chaff which must be ground off. The plant is very hardy, with a short growing season and so is particularly adapted to cold climates and regions such as North Africa where the growing season is limited by lack of water rather than temperature. Barley is even more extensively cultivated by the native populations than hard wheat in North Africa; it is found from the Littoral up to the oases. This grain is particularly important in the southern parts of Morocco, Algeria, and Tunisia, for even in these regions with rainfall below that necessary for the successful growth of wheat, barley will mature, although yields are low and show great fluctuations from year to year. It is an important source of food for both man and domestic animals, the latter consuming it either green or as grain. Considerable quantities are used by the non-Moslem population in the production of malt for brewing, and much of the exports were also utilized for that purpose.

Morocco produces large quantities of barley according to climatic conditions. Considerable quantities were grown for export (mostly by the natives), but the Arabs and Berbers utilize it as a stock feed and they themselves do not hesitate to eat it, using it much as hard wheat. Production in 1938 on some 1,716,000 hectares amounted to 108,500 tons, an average of 0.65 tons per hectare. Exports from Morocco varied between 23,000 and 200,000 tons, most of which France imported under a quota system, which assigned 250,000 to 300,000 tons to Morocco.

Algerian acreage is usually estimated at about 1,200,000 hectares, yielding from 600,000 to 1,000,000 tons. Algeria annually exported between 17,500 and 70,000 tons yearly, but under present conditions it is likely that livestock needs and increased beer consumption will account for most of the crop throughout North Africa. Unfavorable weather conditions in 1943 have caused erratic yields, and the barley harvest in North Africa is short.

BARLEY *

	Area in 1,000 hectares	*Production in metric tons*	*Yield per hectare, tons/hec.*
Algeria			
1934	1,267	974,000	.77
1935	1,256	718,000	.57
1936	1,262	641,000	.51
1937	1,252	598,000	.48
1938	1,177	578,000	.50
Morocco			
1934	1,555	1,520,000	.98
1935	1,741	779,000	.45
1936	1,661	1,526,000	.92
1937	1,681	826,000	.43
1938	1,716	1,085,000	.65
Tunisia			
1934	480	150,000	.31
1935	550	310,000	.56
1936	300	75,000	.25
1937	620	200,000	.32
1938	300	100,000	.33

* From *International Yearbook of Agricultural Statistics*, 1938–39.

In Tunisia, barley is grown especially by the natives in the central and southern parts. Formerly extensively cultivated in the north, it has been replaced there by soft wheat. The areas planted and the harvests show considerable variation from year to year, ranging between 300,000 and 600,000 hectares planted and 75,000 to 200,000 tons of grain harvested, an average yield of about 0.3 tons per hectare, less than that of either Morocco or Algeria. Tunisia previously exported from 24,000 to 300,000 tons yearly, much of which was used in the production of malt.

Other cereals are of comparatively little importance in the Maghreb, although OATS are rather extensively grown as a stock food by European farmers, mainly in the Tell regions of Algeria. Production usually amounts to some 100,000 to 150,000 tons.

OATS *

Year	*Algeria*			*Morocco*			*Tunisia*		
	Area in 1,000 hectares	*Production in metric tons*	*Average yield in tons/hectare*	*Area in 1,000 hectares*	*Production in metric tons*	*Average yield in tons/hectare*	*Area in 1,000 hectares*	*Production in metric tons*	*Average yield in tons/hectare*
1934	182	172,000	0.95	27	27,000	1.84	35	20,000	0.75
1935	176	105,000	0.65	28	15,000	0.54	25	18,000	0.72
1936	191	175,000	0.92	33	19,000	0.59	25	10,000	0.40
1937	193	138,000	0.72	42	39,000	0.93	37	28,000	0.77
1938	182	158,000	0.87	48	47,000	0.98	40	30,000	0.75

* From *International Yearbook of Agricultural Statistics*, 1938–39.

MAIZE or CORN, a native of the New World, is grown very little, except in Morocco, because of its more exacting water and cultural requirements. It is rather extensively cultivated, either under irrigation or dry-farming, in the coastal regions of Morocco from Rabat southward to Casablanca, Mazagan, and Safi (where the heavy fogs produced by the cold Canary Current enable it to thrive in spite of the rather low rainfall) and in the region to the south of Meknès along the foot of the Atlas ranges. It is also grown under irrigation as far south as the Sous Valley, where two crops are obtained yearly. Since the establishment of the Protectorate, the production of maize by the natives has increased; the French, however, have generally found it unprofitable. Some 450,000 hectares (of which not over 15,000 are worked by European farmers) are grown, with an annual yield of 150,000 to 300,000 tons. A portion of this is consumed locally in place of hard wheat or barley; as a stock food it is becoming more important. The remainder was exported, amounting in some years to 60,000 tons, most of which went to France. Exports in 1938 were, however, almost nil and 20,390 tons were imported. Algeria produces small quantities and in Tunisia it is confined to the comparatively well-watered northern portions or to irrigated land. Both Algeria and Tunisia annually imported considerable quantities of corn.

MAIZE *

Year	*Algeria*		*Morocco*		*Tunisia*	
	Area in 1,000 hectares	*Production in metric tons*	*Area*	*Production*	*Area*	*Production*
1934	8	7,000	399	246,000	25	7,000
1935	6	4,000	388	139,000	17	6,000
1936	6	3,500	455	301,000	18	3,000
1937	6	3,000	453	161,000	27	6,000
1938	6	4,000	462	217,000	17	5,000

* From *International Yearbook of Agricultural Statistics*, 1938–39.

GUINEACORN or SORGHUM, called "Bishna" by the Arabs, is particularly adapted to North African agricultural conditions. The root system is rather extensive and the leaves have the tendency to curl up in dry weather. In addition, some strains mature in two months so that the spring rains furnish sufficient moisture for its growth. Throughout the great Savannah regions of West Africa almost to the edge of the Sahara, this small grain is an extremely important food crop (see previous chapters), and it seems somewhat strange that it is not cultivated more extensively to the north of the Desert. It is a traditional crop in North Africa, but many of the Arabs do not seem to regard it very highly as a source of food, relying on it only in the absence of other grains. In Morocco some 85,000 hectares yield 30,000 tons of grain.

The pith of some varieties has a high sugar content and the juice when boiled makes sorghum molasses or syrup. Since methods of crystallizing sucrose (cane sugar) from sorghum syrup have been developed, it has been suggested that increased production of sweet sorghum is perhaps one answer to the sugar deficiency of the Maghreb. Broom corn, the variety *technicum*, is cultivated to some extent; 1,600 hectares were grown in Morocco in 1936.

RYE is cultivated very little in North Africa, although it too is well adapted to conditions there. A few European farmers grow limited quantities in Algeria and Morocco, and in the Rif Atlas and elsewhere it is frequently an important item in the Berber diet. The Arab populations are prejudiced against the grain (for no apparent reason), and the efforts of the French to introduce it have not met with much success.

RYE *

Year	*Algeria*			*Morocco*		
	Area in 1,000 hectares	*Production in metric tons*	*Yield in tons per hectare*	*Area in 1,000 hectares*	*Production in tons*	*Yield in tons per hectare*
1934	1	1,100	.87	1	1,600	.59
1935	1	400	.43	2	600	.28
1936	1	700	.50	3	900	.32
1937	1	900	.73	4	700	.19
1938	2	1,100	.60	3	1,100	.33

* From *International Yearbook of Agricultural Statistics*, 1938–39.

LEGUMES

Next to the cereals, various types of legumes are of greatest importance as a source of energy in the North African diet. Broad beans, horse beans, kidney beans, lentils, and chick peas are all grown, and although acreages are rather small form an important item, some type or other being cultivated almost everywhere. Most legumes require from 15 to 20 inches of rainfall annually, but several types are particularly adapted to arid and semiarid areas.

BROAD BEANS cover a total area of approximately 30,000 hectares in Algeria, producing between 15,000 to 20,000 tons annually, depending on rainfall. In Tunisia some 20,000 to 25,000 hectares yield a like tonnage yearly, while in Morocco 50,000 to 65,000 acres produce 17,000 to 21,000 tons of beans. Practically all of these are consumed within the country.

DRIED KIDNEY BEANS are rather extensively used, and in Algeria this type of legume occupies between 400 and 500 hectares of land, with yields of about 300 tons of dried beans annually. Moroccan acreage ranges between 1,000 (in 1937) and 700 hectares (in 1938), with a production approximately equal to that of Algeria.

LENTILS, one of the most ancient of food plants as well as one of the most nutritious, are rather widely grown in Algeria and Morocco.

The plant has the advantage of being able to resist temperatures as low as 22° F. and for this reason is particularly adapted to continental regions with wide extremes of temperature. In Morocco, 15,000 hectares were planted in 1938, yielding 4,100 tons of the small, lens-shaped seeds, while in Algeria in the same year 2,200 hectares produced 1,000 tons of lentils. The lentil is not widely cultivated in Tunisia but is well suited to that area and can be grown in southern Tunisia, on the island of Djerba and even at Matmata, to the east of the Chott Djerid, when sown in December or January.

CHICK PEAS, native in southern Europe, are extensively grown around the Mediterranean Basin. The plant is a branching, deep-rooted annual which matures in 90 days, and which is well suited to arid and semiarid regions. Unfortunately the foliage is poisonous so that it cannot be used for forage. The green pods are sometimes eaten, however, and the nutritious seed is widely used in soups. Some 40,000 hectares are planted annually in Morocco, with a yield of from 14,000 to 19,000 tons of peas, all of which are consumed within the country. In Algeria, 15,000 hectares produce 7,500 to 8,000 tons, and in Tunisia annual production is usually about 4,000 tons, with areas varying between 5,000 and 12,000 hectares.

SOYBEANS are cultivated to some extent in North Africa, and under a German plan for production of vegetable oils increased acreages were planned for the 1943 season in Morocco and Algeria, in the latter amounting to 1000 hectares.

The CAROB, a small evergreen tree native in the Mediterranean region, is cultivated for its long pods and seeds which are a valuable cattle food. In Algeria there are at least 1,500 hectares of carob orchards and some 250,000 scattered trees. Between 10,000 and 15,000 tons were exported annually during the past ten years. In Tunisia there are some 20,000 trees in production, and the harvest amounts to about 1,500 tons.

FENUGREEK, an annual legume with white flowers and long, slender pods, is grown mainly for forage in North Africa, although the small aromatic seeds are used to some extent as a spice. In Morocco, 4,000 to 5,000 hectares are planted, with a yield of from 1,000 to 2,000 tons. Other types of legumes, such as BERSIM, ALFALFA, and VETCH, are frequently planted with fenugreek or barley or oats, and these together are an important source of hay in all three North African territories.

ROOT CROPS

WHITE POTATOES are the only important root crop in North Africa. Algeria and Tunisia are particularly important producers of both summer potatoes and new potatoes for export. Planted in the fall and grown during winter and spring, the crop matures in spring or summer in the Tell regions of Algeria and in the western Littoral and Sahel of Tunisia. In Algeria approximately 17,000 hectares are planted in potatoes, with an annual production of 145,000 metric tons. Tunisian areas range between 2,000 and 3,000 hectares, with an annual production of 6,000 to 10,000 tons. In the Tunis market in 1938 there were 6,692 tons of Tunisian potatoes as against 986 tons of imported potatoes. Moroccan production of new potatoes for export is about 7,000 to 10,000 tons. (See also Vegetables, page 89.)

The TIGER NUT or EARTH ALMOND, a native of the Mediterranean region, producing small nut-like tubers, is more important in North Africa in its role of a weed than in that of a food, although it is occasionally eaten. (See preceding chapters.) The plant is now cosmopolitan and in North Africa as in the southeastern United States is an important pest in cultivated ground, especially citrus orchards.

VINEYARDS

The WINE GRAPE, the "vine" of the ancients, has been cultivated in Mediterranean countries from time immemorial. It is thought to have originated in the region of the Caspian Sea but has been grown in Egypt for 6,000 years. Much esteemed by the Greeks and Romans, it spread over most of Europe with Roman civilization, and varieties are now found in all temperate regions. The fruit is rather fleshy with a high sugar content and is used for wine, raisins, and as a table grape.

In North Africa, and in Algeria in particular, the great vineyards are of paramount importance. The vine is in Algeria the most important of the European crops, and wine ordinarily accounts for a large proportion of Algerian exports. The development of the Algerian vineyards since 1880 is said to be the principal economic feat of the present period. From 20,000 hectares in 1880 to 123,000 in 1896, the vineyards have grown to their present 400,000 acres in the Departments of Oran, Algiers, and Constantine. This phenomenal growth of the Algerian vineyard began with the wine crisis in France in the late eighteen-seventies. By that time the root louse (*Phylloxera vitifolii*), discovered on wild grapes in eastern North America in 1854, had reached the European continent and devastated much of the French vineyard. It is said that land worth 8,000 to 16,000 francs per hectare sold for 1,000 francs after being attacked by phylloxera. Under these conditions the vineyard owners of the Midi immigrated in large numbers to Algeria, establishing new vineyards in the phylloxera-free Tell regions.

By 1889, 1,500,000 of the 2,500,000 hectares of vines in France had been lost, and production was short by an estimated 60,000,000 hectoliters of wine.* The root louse reached North Africa in its turn, first in 1885 at Tlemcen, then at adjacent Sidi-bel-Abbès, later at Philippeville and Oran; but, by the time any real damage had been done, the greatly expanded Algerian vineyards were in production and competition with the French. The French vineyards were, however, reconditioned with grafted vines—wine grape on stocks of native American grapes which are much more resistant to phylloxera—and by 1900, a year of exceptional French production, Algeria was overproducing with consequent serious economic disturbances. Since that time the absorption of North African wine by the French market and increased consumption have made possible further growth of the North African vineyard, which is at present, however, developed to a point where further extension would be dangerous. It is almost certain, nevertheless, that the European wine crisis contributed more to the European colonization of Algeria than any sort of political propaganda.

In Algeria the vineyards are concentrated in the Tell regions and in the Tell Atlas; the vine-growing zone for the most part does not extend inland more than 75 miles. Beyond this the winters are too cold and the vines are exposed to the excessively dry blasts of the sirocco or guebli (a hot dry wind blowing from the Sahara regions in the summer months). The greatest concentration of vineyards in Algeria is in the Department of Oran, where some 240,000 hectares are cultivated. The Department of Algiers is second, with about 125,000 hectares, and the Department of Constantine has only 26,000 hectares, concentrated mostly in the regions around Bône and Philippeville. This difference in acreage between the three departments is directly proportional to the density of the rural European populations, for wine production is almost exclusively European since the use of wine is forbidden to Mohammedans by the Koran.

In Oran, the main producing areas are in the vicinity of Oran, Mostaganem, Mascara, Tlemcen, Relizane, Aïr Temouchent, and Sidi-bel-Abbès. The dry climate there is particularly favorable in that it restricts the development of fungal diseases. Of the 240,000-odd hectares composing the Oran vineyard, about 92,000 are of

* 1 hectoliter = 26½ gallons approximately.

ungrafted vines; the rest were reconstituted with grafted vines following the appearance of phylloxera in those regions. At Relizane, after phylloxera appeared, the vineyards were destroyed and for fifteen years only cereals were grown. After a rise in prices on the French market, they were replanted with ungrafted vines and interplanted with fruit trees—citrus, apricots, and olives—so that should the vines be again destroyed by phylloxera, the other crops would furnish a source of revenue, preventing total ruin.

In the Department of Algiers, the main vine areas are within a relatively short radius of Algiers, in the sheltered valleys back of the coast, notably the plain of Mitidja, and in the areas around Miliana and Médéa.

The Tunisian vineyard is nearly all concentrated in the relatively well-watered extreme northeastern portion in the neighborhood of Tunis, Grombalia, Cape Bon, Souk-el-Arba, and Mateur. Although it now covers about 50,000 hectares, at the time of the establishment of the French Protectorate over Tunisia, there were only about 1,100 hectares of native vines, grown only for table grapes. At the present time, Mohammedans own only 4,000 hectares of vines, widely scattered, even as far south as the oases of the Chott Djerid, where in the presence of abundant water the vines clamber up the trunks of date palms.

Most of Tunisia has been spared the devastations of the phylloxera, but it appears that in spite of rigorous control measures the insect is gradually spreading. It first appeared in Tunisia at Souk-el-Khemis in 1906. The vineyards there were destroyed, but in 1928 the insect was found at Le Kef and in 1936 at Nassen, less than 10 kilometers from Tunis in the heart of the grape-growing region. Although all the infected vineyards have been destroyed, and up to the beginning of the war, at least, strict inspection of all vineyards was practised, it is evident that the Tunisian vineyard must eventually be reconstituted with grafted vines.

In Morocco the native populations are growing about 20,000,000 vines totaling some 19,000 or 20,000 hectares. European vineyards cover about 23,000 to 25,000 hectares, mostly in the rolling country around Meknès, in the hinterlands of Casablanca and Mazagan (the Chaouïa), and in the Triffas Plain in eastern Morocco at the mouth of the Oued Moulouya.

Wine production varies from year to year, but North Africa has an annual exportable surplus of from 15,000,000 to 20,000,000 hectoliters. Algerian production is ordinarily in the neighborhood of 20,000,000 hectoliters (maximum 22,000,000 in 1934 and minimum since 1900, 4,500,000 in 1902). Tunisian production is usually about 1,500,000 hectoliters (125,000 in 1915 and 1,976,000 in 1938). Moroccan yields amount to about 960,000 hectoliters (1938), of which about 450,000 are consumed within the country. These figures contrast with a French production of 61,000,000 in 1938 and an Italian production of 34,000,000 in 1937, the last years for which figures are available.

Ordinary wines, dry and sweet wines, liqueurs, and mistelles are produced and are said to compare well with the French products, although North African producers have concentrated on quantity rather than quality. Large and modern wineries are scattered throughout the vineyard areas, and fermentation processes are carefully controlled. The Tunisian liqueur, byrsa, is unique.

It is reported that before the outbreak of the war efforts were being made to increase the consumption of grape juice and to utilize grapes in other ways. A large part of the wine production (perhaps one-half) was distilled into alcohol to help alleviate fuel shortages following the German conquest of France. How much of this went to Germany is not certain. In addition to wines, Morocco annually produced from 1,500 to 2,500 tons of raisins, and table grapes from native vines amounted to some 26,200 tons in 1937, making them an important fresh fruit among the Mohammedan populations at least. In Algeria the export of early table grapes was an important item, amounting to 7,000 tons in 1937 and 11,800 in 1938. Tunisian production of table grapes from Mohammedan vines amounts to 6,000 to 15,000 tons annually.

It is certain that the North African vineyard has been overproducing. High tariffs have in the past to a large extent prohibited the export of wine to foreign countries, and France absorbed most of the excess. As a result, the French harvest determined the price received for North African wine. Exceptional yields in France spelled ruin for the North African "monoculturists," and it is now clear that there is a permanent and lasting over-production in the latter region as long as French production continues. As a step toward alleviating the situation, the Moroccan government in 1935, at the demand of France, forbade the planting of any more vineyards in spite of the resistance of the producers and the chambers of agriculture. Further planting of vineyards has also been prohibited in Algeria and Tunisia, and at least 16,000 hectares have been dug up in Algeria and 9,000 in Tunisia. The unfortunate thing about viticulture in North Africa is that it is a monoculture, although vines are in a few instances planted with fruit trees, and that it produces a product which is not thought of as a necessity except in southern Europe. In addition it has always been in direct competition with the great French wine industry. At the present time North African producers are probably unable to dispose of their product both because of the lack of shipping facilities for wine and the lack of coal for distillation of wine into alcohol.

NORTH AFRICAN VINEYARDS

	Area of all vineyards (in 1,000 hectares)	*Area of vineyards for wine (in 1,000 hectares)*	*Wine production (1,000 hectolitres)*	*Vineyards for table grapes*		
				Area	*Production (metric tons)*	
					Total	*Exports*
Algeria						
1928–32	345					
1934	405	388	22,043	5		7,600
1935	408	400	18,910	4		11,500
1936	396	390	11,527	5		6,100
1937	408	395	15,424	5	12,600	7,000
1938		399	21,490	4	15,700	11,800
Morocco						
Europeans						
1928–32	11					
1934	23		580			
1935	26		520			
1936	25		265			
1937	24		582			
1938	24		748			
Natives						
1928–32	15[a]					
1934	17			17[a]	27,000	
1935	18			17	17,000	
1936	19			18	15,000	
1937	20			19	26,200	
1938	20					
Tunisia						
1928–32	43					
1934	54	44	1,700	4	6,000	
1935	49	45	1,700	4	6,500	
1936	47	43	1,420	4	5,500	
1937	47	43	1,454	4	15,000	
1938	46	42	1,976	4	15,000	

[a] Million plants. (1,000 ha. = 2 million plants.)

OIL PLANTS

Throughout all of North Africa, the olive is the main source of edible oil, especially among the native populations. The total production is not so large as that of Spain, but the region was before the war by far the leading exporter of olive oil; 74,420 tons were exported in 1938, an exceptionally good year. Of this quantity, Tunisia, the most important producer, exported 40,480 tons; Algeria, 24,380 tons; and Morocco, which is often deficient, 9,560 tons. Much olive oil is consumed locally, and Tunisia is usually the only region with an important surplus.

In southwestern Morocco, oil obtained from the fruit of the unique argan tree largely replaces olive oil, but throughout most of Morocco peanut oil is usually imported from West Africa to make up for a shortage of edible oils.

In addition to edible oils, the region produces small amounts of crops yielding other types. Flax, on the plains of western Morocco, in some years covered as much as 23,000 hectares, depending on the condition of the world market. Linseed from this amounted to 10,000 tons in 1937 and 5,600 in 1938. The castor bean does well in most regions, but the low prices obtained usually discourage the planting of this crop. Cottonseed has been an unimportant source of oil, and soybean and sunflower have been grown to a small extent. After 1940, when oils were strictly rationed in North Africa, most of the production went to Germany which was deficient in fats. Moroccan farmers were ordered to put at least a tenth of their cultivated area under oil-bearing plants. In Algeria increased acreages of soybean, sunflower, flax, cotton, and castor bean were planned for 1943. Even grape seeds were utilized.

Under the more beneficent Allied occupation it should be possible to obtain quantities of olive oil from North Africa, and from Tunisia in particular, for the extensive olive groves suffered very little damage in the battle for Tunisia.

OLIVES. The olive is one of the oldest fruits and has been grown from prehistoric time. It has been cultivated in Egypt from at least the seventh century, B.C., and is widespread around the Mediterranean Basin where it is native. A small evergreen tree, it reaches great age, becoming at length gnarled and twisted. Olive oil is one of the most important of edible oils, particularly in the Mediterranean area and in regions of similar climate. The oil obtained from the ripe fruit, by either hand or mechanical methods, is golden yellow, clear, odorless, and has the advantage that it will keep for a long time, becoming rancid only when exposed to air.

In many areas of North Africa, especially eastern Tunisia and the limy and little-watered lands of Algeria, the land can be fully utilized only by olives. The plant is, however, widespread wherever rainfall and temperatures (an average temperature of 57° F., not falling below 14° F., is desirable) are suitable and still grows wild in some regions.

In Tunisia, the main olive-oil-producing area of North Africa and the third among exporters of olive products, there are some 19,000,000 trees, most of which are owned by Moslems; 220,000 Moslems pay the kanoun tax as compared with 1,000 European owners.

In the northern regions the centers of olive growing are small and scattered, being concentrated mainly around Beja, Bizerte, Tabarka, Souk-el-Arba, Tunis, Zaghouan, Grombalia, Teboursouk, Maktar, and Mejez-el-Bab. Throughout this region annual rainfall varies between 16 and 20 inches so that conditions for the cultivation of the 4,000,000 olive trees there are more favorable than in other areas.

The Sahel and central region, including the Fourth Administrative Area, has some 7,200,000 trees, concentrated particularly on the undulating lands in the neighborhood of Sousse and inland around Kairouan and Thala. Rainfall in this area averages about 12 inches.

In the southern region around Sfax, the olive groves have grown in less than fifty years from 360,000 trees to 7,600,000, covering an area of 450,000 hectares. The main area is around Sfax, with scattered groups of olives at Gafsa, Maknassy, Gabès, Matmata, and Tathuine, with fairly extensive groves on the Island of Djerba. The low and variable rainfall permits only 16 to 20 trees per acre (as compared with 630 per acre on the high tablelands), and in this region a form of dry-farming is practised so that all possible moisture is conserved. Although yields are low, much better harvests have been obtained in recent years through the education of the natives in the fundamentals of olive growing.

The table below shows the great increase in the Tunisian olive groves since 1892:

Regions	*Number of olive trees*			*Difference between 1882 and 1938*
	Prior to 1882	*In 1934*	*In 1938*	
North	2,400,000	4,119,000	4,200,000	1,800,000
Sahel and Center	4,500,000	7,062,000	7,200,000	2,700,000
South and Extreme South	1,240,000	5,975,000	7,600,000	6,360,000

From *International Review of Agriculture*, 1940.

Great progress has been made in the methods of processing since the early years of the Protectorate, and there are (or were) over 1,500 oil factories which turn out pure, pleasant-tasting oils. There are still a few of the primitive native maasra in outlying districts which manufacture oils for local consumption only. In addition, there are nineteen factories for extracting oil from olive cake, and a similar number of refineries and soap factories which can process all the olive cake produced by oil extraction plants.

Yields are notoriously variable from year to year, depending upon rainfall, climatic conditions at such critical times as flowering, setting of fruit, etc., and the incidence of insect pests. Tunisian production ranges between 15,000 and 60,000 tons while exports vary likewise.

In Algeria, more attention has been given to vineyards and cereals than to olives, although there are between 9,000,000 and 10,000,000 trees in the region. More than half of these (5,400,000) are concentrated in the Department of Constantine, distributed through the valleys of the Sahel, the Seybouse, at Philippeville, and at scattered areas at Batna, La Calle, Guelma, and Tebessa. In the Department of Algiers there are about 3,100,000 throughout the Kabylia regions and in particular at Tizi-Ouzou. These extensive groves play an important part in the economy of the Berber tribes of the Kabylia. In the Department of Oran there are over 1,300,000 trees, mainly around Mostaganem, Saint-Denis-du-Sig, Mascara, Relizane, and especially in the neighborhood of Tlemcen.

Olive oil yields in Algeria are again very variable, oscillating between 9,000 and 30,000 hectoliters. Much of this production is used locally; the natives know no other edible fat, and the European populations originally from southern Europe are large consumers. It is estimated that the annual consumption of one Kabyle is 15 liters, that of a European 12, and that of an Arab 2.5. Algeria seldom has important quantities for export.

In Morocco, the main producing areas are around and north of Fez (2,600,000 trees), Meknès (400,000), and Marrakech (2,600,000), where the trees are irrigated. There are also irrigated groves in the Sous Valley of southwestern Morocco. Of an estimated total of 7,600,000 trees, about 6,700,000 are bearing and of these over half belong to natives. Native care of the trees is poor, and harvests are correspondingly low. Annual production ranges between 7,000 and 10,000 tons. Morocco is often deficient in olive oil and imports considerable quantities of vegetable oils.

Spanish Morocco has an estimated 320,000 olive trees.

The Olive in North Africa *

	Total no. of trees	Trees in production	Production of olives in metric tons		Production of olive oil in metric tons		Exports of olive oil in metric tons
			For oil	For food			
Algeria							
1933–34	68,000[a]		72,000	9,700	8,300		—[b]
1934–35	69,000		114,000	8,400	14,700		6,560
1935–36	71,000	8,493,000	102,800	7,900	13,000		26,150
1936–37	82,000	8,141,000	54,500	8,100	6,600		11,780
1937–38	81,000	8,239,000	128,600	9,500	16,000		8,680
1938–39					5,900		24,380
Morocco							
1933–34			49,000		7,300		0
1934–35			75,000		11,500		0
1935–36			60,000		9,000		200
1936–37	7,400,000	6,687,000	68,000		10,500		510
1937–38	7,657,000	6,749,000					7,270
1938–39	7,929,000						9,560
Tunisia							
1933–34					60,000	6,000[c]	62,060
1934–35					55,000	5,500	53,810
1935–36					60,000	6,000	49,180
1936–37	17,150,000	12,307,000			15,000	1,500	31,120
1937–38	17,366,000	13,282,000			50,000	5,000	18,790
1938–39	17,766,000	13,283,000			30,000	3,000	40,480

[a] Hectares relating only to dense plantations (comprising somewhat more than half of the trees cultivated).
[b] Exports for the calendar years 1933–38.
[c] Black olive oil.
* Figures from *International Yearbook of Agricultural Statistics*, 1938–39.
N.B. Production figures are undoubtedly underestimated, for export figures exceed production and the local consumption of olive oil is large.

Argan Oil. In southwestern Morocco in the region between the mountains, ocean, and desert from Mogador to Agadir and southward, over an area of about 450,000 hectares, the peculiar argan tree (*Argania sideroxylon*), a spiny evergreen tree in general aspect resembling the olive, forms the dominant vegetation. The area has the aspect of a vast natural orchard, but an orchard in which every tree is different in appearance. In the dry, mild climate of this region, the tree is extremely slow growing and some specimens are known to be over four hundred years old. The argan is resistant to destruction by goats by virtue of its spines, but the young shoots form almost the only forage available at some periods of the year and are eaten by the black Moroccan goats which climb up into the trees. The fruit is a drupe (stone fruit) similar to the olive but with an extremely unpleasant taste which is, however, relished by the goats. The seeds or pits contain an oil somewhat like olive oil, which is extracted by the natives of the region and which is highly esteemed, although repugnant to the European palate. Comparatively small quantities of argan oil are produced, but in this region of southwestern Morocco the native economy is to a large extent based on the argan. It does not seem likely that any other crop will supplant this peculiar tree indigenous to these semidesert regions.

FRUIT TREES

The culture of citrus fruits in North Africa is confined mostly to the coastal areas. Farther inland the dangers of damage by frost and sirocco are too great. Citrus species, most of which are indigenous to humid, southeastern Asia, must be grown with irrigation in the Maghreb, and the distribution of citrus plantations is limited, in part at least, by the available water. ORANGES of various types, MANDARINS, including a large percentage of TANGERINES and CLEMENTINES,[1] LEMONS, LIMES, CITRONS, BITTER ORANGES, GRAPEFRUIT, and KUMQUATS are all grown, but oranges, tangerines, and lemons are most important.

Citrus fruits in Algeria are confined mostly to the coastal plains and are concentrated in the plain of Mitidja near Algiers, in the Bône district, in the region of Philippeville, in the valley of the Oued Sig in western Algeria, and in the regions around Perrégaux, Relizane, and Ferme Blanche in the valley of the Chélif or its tributaries where abundant water is available for irrigation.

In 1938, 11,300 hectares were planted in citrus, and of this area oranges occupied 6,300; clementines and mandarins, 4,500; and lemons, 500. Production is usually about 100,000 tons yearly, divided roughly 50, 45, and 5 per cent between oranges, mandarins, and lemons. Exports in the past represented about 70 per cent of production, or 70,000 tons, about half oranges and half mandarins.

[1] Tangerines were so named because they were first shipped to England from Tangier. A seedless variety developed by Père Clément about 1900 at Oran is widely known by the name "clementine" in Europe and North Africa.

Practically all of this export was absorbed by France during the season from October to May, and had a value of 100,000,000 francs.

In Morocco, the chief producing centers in order of importance are Meknès, Rabat, the Gharb region of northwestern Morocco, the Zegzel Valley of eastern Morocco, Casablanca, its hinterland the Chaouïa, Marrakech, and Fez. It is likely that the Meknès region will become the important center of orange production in Morocco, for in this region conditions for citrus are very favorable, there being a total absence of frosts, a supply of fresh water, and a dry climate which checks the development of fungal diseases and scale insects, the worst enemies of citrus fruits. In Morocco, the orange was widely distributed before the coming of the French, and native types have been developed which are fairly hardy, spreading from the coast up to altitudes of 2,500 feet in some sheltered valleys in eastern Morocco. Native groves throughout the country amount to about 1,800 hectares or 520,000 trees. In 1938, the European groves covered 5,614 hectares, of which 3,300 were under oranges, 1,700 clementines, 450 mandarins, 119 lemons, 25 grapefruit, and the remaining 20 under other types of citrus fruit. It should be noted, however, that a large part of the citrus plantations in Morocco have not yet come into production.

Annual production in Morocco averages 35,000 tons, of which about 25,000 are consumed locally and the rest exported. Local consumption could probably be increased to 50,000 tons per year. Exports from Morocco increased continually in the last ten years. Practically all of these went to France. In 1930, the total exports amounted to about 300 tons per year while recent figures note approximately 12,500 tons composed of 77 per cent oranges, 20 per cent mandarins, and 3 per cent lemons.

In Tunisia, citrus cultivation is not as important as in Algeria and Morocco where most of the plantations are owned by Europeans. In Tunisia, European farmers have devoted their attention to cereals, grapes, and olives, so that over 80 per cent of the Tunisian citrus orchards are under native cultivation. In 1938, citrus plantations covered about 2,000 hectares, with oranges occupying 67 per cent (1,347 hectares) of this area. Lemons were on 300 hectares and clementines and mandarins 280 hectares.

Two-thirds of the citrus plantations are in the Cape Bon region, while a fourth of the plantations are in the neighborhood of Tunis. The remainder are scattered groves in the regions of Bizerte, Sousse, and points in the southern oases. Lack of water for irrigation is one of the chief limiting factors in the expansion of the citrus orchards in Tunisia. Annual production is about 15,000 tons, of which 80 per cent is oranges. Approximately 8,000 tons are consumed locally, but this is undoubtedly subject to great increase in the future. The high prices obtained on the French market after the Spanish Civil War had stopped most of the citrus exports from Spain, encouraged production in Tunisia, and practically all of Tunisian exports went to France. These amounted to 7,000 tons annually, 70 per cent oranges, 23 per cent mandarins, and 7 per cent lemons.

The North African areas supplied France with 25 per cent of her total imports in oranges, 92 per cent in mandarins, and 8 per cent in lemons. Grapefruit production in North Africa is negligible, and some 6,000 tons for the French market came from Palestine.

Other types of fruit trees are many and varied and are widely grown. Particularly important are almonds, apricots, figs, pears, peaches, cherries, plums, and pomegranates. Bananas are grown in sheltered coastal areas and in some of the oases along the edge of the desert, and guavas, cherimoyas, loquats, tamarinds, and jujubes are cultivated to a small extent in sheltered places along the Littoral.

The ALMOND, closely related to the peach which it resembles in flower and young fruit, is widely cultivated throughout North Africa where it permits the utilization of many lands otherwise too dry. In Tunisia there are some 800,000 bearing trees which produce from 2,000 to 4,000 tons of nuts. Local consumption is rather high and only about 450 tons were exported yearly. In Algeria, there are 1,800 hectares of almonds in orchards and around 300,000 scattered trees. Exports amounted to 300 to 400 tons. Dried almonds are produced particularly in the regions of Miliana, Médéa, and Aïn-Bessem in the Department of Algiers, and in the Department of Oran. In Morocco, particularly in the southern areas along the foothills of the Atlas ranges, the almond is an important item of commerce. Oujda in eastern Morocco, the central Atlas, Mogador, and Marrakech are the most important regions, although the almond is grown almost everywhere. Morocco has about 4,500,000 trees, and the culture could be further extended. Exports amounted to 1,400 tons in 1937 and 1,800 tons in 1938. Yields are, of course, variable from year to year according to climatic conditions. Algerian and Tunisian almonds ripen before those of Spain and Italy, while the Moroccan fruit matures still earlier. Prior to the outbreak of hostilities, fresh almonds were exported to France. Spain and Italy are the world's greatest almond producers; the United States is third, while Morocco is fourth, Algeria eighth, and Tunisia tenth.

APRICOTS are widely cultivated, and in Tunisia some 300,000 trees annually produce 50,000 tons of fruit. In Algeria, 828 hectares of apricot orchards are planted, mainly in well-watered regions, and there are 130,000 scattered trees. The high and relatively well-watered Aurès ranges near Batna are one of the centers of apricot production. Dried apricots are produced on a small scale. Apricots are also cultivated in Morocco, but to what extent is not certain.

THE FIG is found nearly everywhere in North Africa, where it should be counted more as a staple food than as a fruit. Particularly in the mountainous Kabylia regions of Algeria, between Algiers and Bougie, it is an important food. It is only by virtue of the plantations of olives and figs that the Kabylia can support the dense population of Berbers that it does. (A density of 500 persons per square mile is not infrequent!) There are an estimated 7,000,000 trees in Algeria, both in plantations and in scattered stands. About 25,000 tons are produced annually for consumption in the fresh state, while some 40,000 to 55,000 tons are dried, the production of dried figs being estimated between 15,000 and 20,000 tons. Annual exports range between 5,000 and 10,000 tons. The best qualities are eaten, but the inferior products were utilized principally in Central Europe in the preparation of a coffee substitute. In Morocco, the fig is scarcely less important, and there are estimated to be about as many trees there as in Algeria. Figs in Tunisia likewise play an important part in the native diet and are consumed mostly within the country. There are an estimated 1,000,000 trees producing 15,000 tons of fruit. The culture of the fig is essentially a family occupation, and methods of drying the fruit are for the most part primitive.

CHERRIES, PLUMS, POMEGRANATES, and PEACHES are grown rather extensively in many fairly well-watered regions through the Algerian and Tunisian Tell and in northern Morocco, and early varieties were exported to Europe as "primeurs." The native populations consume nearly the entire crop of POMEGRANATES, of which many varieties are known. In Tunisia, this small tree is usually grown with oranges in the region of Cape Bon and Bizerte. The MEDLAR, a small tree with a brown, apple-shaped, acid fruit, is cultivated to some extent as is the QUINCE. The APPLE and PEAR are not particularly well suited to North African conditions, although species of pear are native to the region. CHESTNUTS are grown in Morocco, mainly by Berbers on the upper slopes of the Middle Atlas ranges.

WALNUTS are found in both the Rif and Middle Atlas, and are utilized to some extent by the Berber populations. The acorns of the EVERGREEN OAK (*Quercus ilex*) and the CORK OAK (*Quercus suber*) are edible and sometimes are an important food for natives. BANANAS are cultivated in Tunisia in the oasis of Gabès, and scattered groves exist along the Algerian Littoral in areas free from frost and with sufficient water. In Morocco, the plant is grown under irrigation in the Sous region at the edge of the desert, but the culture is not very extensive and might well be replaced by sugar cane or some other more practicable crop.

MINOR FRUITS IN NORTH AFRICA *

	Algeria	*Morocco*	*Tunisia*
Apple and Pear	1,797 hectares and 97,000 scattered trees	3,760,000 trees	400,000 trees
Cherry	931 hectares and 54,000 scattered trees	190,400[a]	?
Medlar	1,979 hectares and 97,000 trees	?	?
Peach	?	?	100,000
Pomegranate		498,000	800,000
Plum	2,400 hectares and 180,000 trees	?	50,000
Quince	?	?	150,000

* Figures for Tunisia and Morocco for 1936. Algerian figures from the quinquennial statistics last published for 1930.

[a] Includes chestnuts.

The DATE PALM has been cultivated in the Near East for at least 5,000 years and in the desert regions has long been the staple product and almost the only wealth of which the oases could boast. It is probably a native of India or Arabia, but reached Egypt before the Christian era and was introduced into the Sahara region by invading tribes from the East. The date will grow in North Africa as far north as Tangier, but it is only in the oases of the Sahara that it reaches full development and only there that it fruits. Contrary to popular conception, the date palm needs abundant water but at the same time requires a hot, dry climate for fruiting. Male and female flowers are on different trees and in cultivation 90 per cent of the male trees are removed. The numerous female flowers, sometimes 10,000 to a cluster, are always pollinated by hand, a practice in use since at least 2400 B.C. Clusters of male flowers for pollen are sold in the native markets in April. In addition to the fruit, the growing tip of the palm itself is sometimes eaten, but this kills the tree. The sap is rich in sugar and is sometimes collected from cuts made near the top of the trunk and fermented to make palm wine.

Algeria is the principal producer of dates in North Africa, with an annual production of about 100,000 tons. Of these, a limited quantity (13,000 tons) are of the excellent *deglat nour* (finger of light) type which is harvested from October to December. This is the type which was extensively exported to Europe. The remainder of production, including soft and dry dates (over 100 varieties are grown) is consumed by the native populations of the desert regions.

The 6,800,000 Algerian palms, practically all of which are owned by natives, are divided among four main groups of oases: the Saoura, Gouara, Touat, and Tidikelt oases of western Algeria, extending some 700 miles from the Atlas ranges into the heart of the desert and known to Arabic writers as the "Street of Palms"; the M'zab oases centering about the seven cities of M'zab, of which Ghardaïa is the most important; the Ziban oases of the region of Batna and the oases fed by artesian wells along the Oued Rir south of the Chott Melrir; and the Souf oases of the extreme east of Algeria, centering around El Oued. The western Algerian oases have about 2,000,000 palms scattered along the bed of the Oued Saoura and the line of a great geologic fault which directs seepage water toward that area. This water is collected by a system of underground tunnels known as "foggaras," which extend from the oases in all directions bringing water from underground sources within a radius of several miles. (Similar tunnels extend from the city of Marrakech into the foothills of the High Atlas, supplying water for the irrigation of the fruit orchards in that part of Morocco.) The M'zab oases with about 260,000 palms are located on a plateau in the Sahara almost directly south from Algiers. At this point the water supplies are located beneath 200 feet of hard limestone and must be laboriously raised from deep wells to the surface by donkeys. This method is expensive, and the M'zab oases are not self-supporting but are maintained as a sanctuary by the Mozabites who throughout North Africa are merchants, moneylenders, and bankers.

The rich oases clustered about Biskra, north of the Chott Melrir, are known as the Ziban and are one of the centers for the deglat nour date. Other centers of Algerian production are farther south along the Oued Rir, where artesian wells up to 200 feet deep supply the oases as far as Touggourt and Ouargla. About 3,500,000 date palms are included in these groups. Farther to the east at El Oued, where water supplies are near the surface, each garden is a broad funnel dug in the sand to the water-bearing layer. The main problem there is keeping the sand from caving in and covering the trees.

In Tunisia, the oases are confined to the southern region around the Chott Djerid, the oasis of Gabès, Gafsa, and the Island of Djerba. The Djerid oases on the northwest of the Chott Djerid and the Nefzaoua on the south, both supplied by artesian wells, account for the Tunisian production of deglat nour dates, Tunisia producing about one-sixth of the total for North Africa and Algeria the remainder.

Moroccan dates are generally conceded to be much inferior to those of Algeria and Tunisia, and although there are large oases as far north as Marrakech, the dates there are little esteemed. It is estimated that there are about 1,000,000 date palms in Morocco, but the division among the oases is more or less problematical and production is variable. The main groves are at Marrakech in the dry plains just north of the High Atlas, in the Sous Valley between the Anti-Atlas and High Atlas, in the valley of the Draa, and in the Tafilalet oases watered by the Oued Rheris and the Oued Ziz. The oases along the Oued Guir in the extreme east are also of some importance, as is Figuig, although these are geographically more a part of the system of oases of western Algeria.

In connection with date culture it should be noted that small quantities of many other crops are grown under irrigation in the shelter of the palms. Small amounts of cereals are grown in the wintertime; vegetables of various types are an item; and apricots, almonds, figs, and other fruits are grown. The oases of southern Morocco, dependent as they are upon flood waters from the High Atlas ranges, are subject to drought; and only since the severe droughts of 1935–37, after which the Moroccan government instituted a series of irrigation and water-storage projects in this region, has there been any real stability.

The principal groups of oases in Algeria and Tunisia with their approximate quotas of date palms are listed below:

Figuig (Morocco)	100,000	Oued Rir	1,700,000
Gourara	757,000	Souf (El Oued)	295,000
Touat	442,000	Ziban	834,000
Tidikelt	365,000	Djerid	900,000
M'zab	260,000	Nefzaoua	280,000
Ouargla	1,000,000	Gabès	200,000

DATES IN NORTH AFRICA *

	Algeria		Tunisia		Morocco	
	No. trees	*Production in metric tons*	*No.*	*Production in metric tons*	*No.*	*Production in metric tons*
Deglat nour	844,000		194,000		none	
(bearing)	726,000	13,300	92,000	4,000		
Common dates	5,960,000	92,300	2,481,000			
(bearing)	4,554,000		1,414,000	28,000		
Soft dates	3,520,000					
(bearing)	2,886,000	53,800				
Dry dates	2,440,000					
(bearing)	1,668,000	38,500				
Total	6,804,000	105,600	2,675,000	32,000	1,047,000	?

* Figures for the year 1937, from the *International Yearbook of Agricultural Statistics*, 1938–39.

VEGETABLES

Under irrigation, practically any type of vegetable of temperate or subtropical climates can be grown in North Africa. Indeed, practically all of the common garden vegetables and fruits of Europe and America are cultivated in small irrigated patches almost everywhere by both native and European alike. Among the more important are TOMATOES, GLOBE ARTICHOKES, GREEN BEANS, GARDEN PEAS, CABBAGE, LETTUCE, CELERY, FENNEL, TURNIPS, CARROTS, ASPARAGUS, SQUASHES, and PEPPERS. The ONION is everywhere, but local production ordinarily does not suffice and quantities are imported. GARLIC is frequently grown. WATERMELONS, CANTALOUPES, and other melons do well and are much appreciated during the dry summers. (In southern Morocco the young melons are buried and left to mature underground away from the burning rays of the sun.)

Many wild plants are utilized by the natives, particularly members of the families to which spinach, buckwheat, and the amaranth belong. Some wild legumes are eaten, but occasional cases of paralysis caused by eating certain types of sweet pea (*Lathyrus*) are reported. In time of famine, the hearts (growing tips) of the dwarf palm are eaten by the natives as a food, as the palmetto is eaten in the southeastern United States.

The last twenty-five years have seen a rather remarkable development in North Africa of truck gardening for export, especially in the Littoral regions around Algiers, Oran, Philippeville, and Bône in Algeria, and more recently Casablanca, Rabat, and Mazagan in

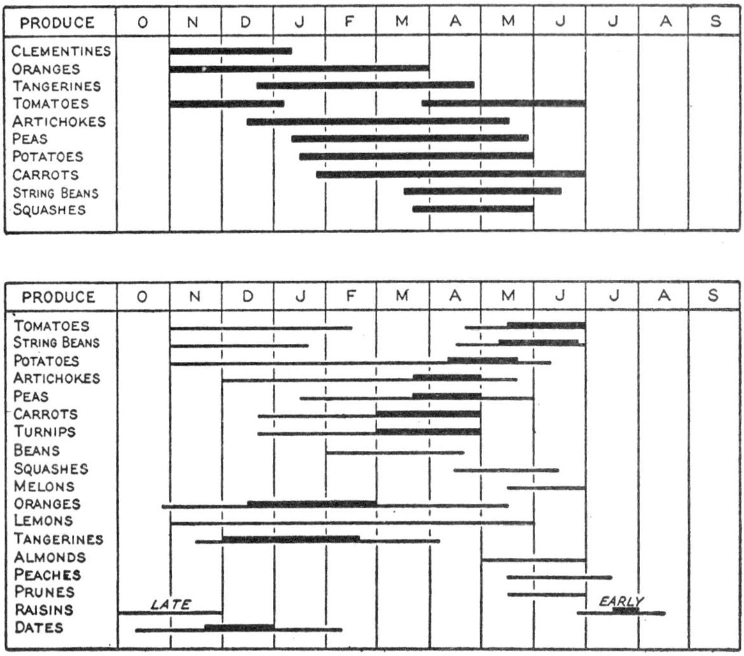

FIG. 24

CALENDAR OF EXPORTS OF EARLY VEGETABLES AND FRUITS

Morocco (*above*); Algeria (*below*)

The heavier line in the lower graph indicates periods of heaviest exports.

Redrawn from *Géographie universelle*, II, Pt. I (1937), 258.

Morocco. The mild winters of the Littoral make possible the culture of a large number of vegetables which prior to the war were shipped through the large North African ports to markets in Europe in fall, winter, spring, and early summer. By virtue of the latitude, Algerian truck crops were ready for European markets long before those from Spain and southern France, while in Morocco the harvest is about two weeks ahead of Algeria. The figure below indicates the main types of vegetables exported as well as the chief periods of export.

Some 39,000 hectares in Algeria are devoted to this type of truck gardening, mainly in the regions of Algiers, Oran, Philippeville, and Bône. Some scattered localities are particularly favored and are able to grow all types of truck crops. In the Department of Algiers they are particularly abundant. Conspicuous examples are Guyotville, Staoueli, and Zeralda. Exports of market garden crops from Algeria amounted to 139,000 metric tons in 1936, nearly half of which was new potatoes raised from imported seed.

In Tunisia, truck gardening is concentrated around Tunis and Bizerte where there are numerous European truck gardeners. In other places it is mainly in the hands of natives. Several areas are specialized in one particular crop: Porto Farina, potatoes; Cape Bon, tomatoes and peppers; the oases of southern Tunisia, asparagus. Tunisian exports are small in comparison with those of Algeria and Morocco. Shipping dates are approximately the same as for Algeria.

Production of truck crops in Morocco has taken a startling jump in recent years (see table) and now centers in the regions around the

Exports of Truck Crops from North Africa *
(In metric tons)

	1921	*1930*	*1934*	*1936*
Algeria	19,300	64,300	81,587	138,923
Morocco	2	4,200	21,582	26,220
Tunisia	4	140	1,256	5,400
North Africa	19,306	68,640	104,425	170,543

* From *La Tunisie agricole*, 39: 51, 1938.

Exports of Various Types of Truck Crops in 1936 *
(In metric tons)

	Algeria	*Morocco*
New potatoes	72,415	7,000
Tomatoes	17,546	13,000
Artichokes	22,062	2,600
Garden peas	4,082	200
Green beans	8,421	1,600
Carrots	13,027	600
Various	1,370	1,200
	138,923	26,200

* From *La Tunisie agricole*, 39: 54, 1938.

large ports where shipping facilities are available. Of particular importance is the area around Casablanca, where in the summer truck crops are raised under irrigation. In the regions of Rabat-Salé, Mazagan, Port Lyautey (Kénitra), and Ouezzan, truck gardening is also well developed.

Large quantities of fresh fruits and vegetables undoubtedly went to Nazi Germany through Vichy France before the Allied invasion, and it seems that some should soon be available for shipment to England now that sufficient amounts of seed have reached North Africa from the United States. Unfortunately, the facilities for dehydrating fruits and vegetables are limited and far from adequate.

HERBS AND SPICES

The Mediterranean Basin has long been a center for the cultivation of many of the better-known plants used for flavoring food. Some of these have been cultivated for so long that their origin is completely lost, but numerous types are known to be natives of the Mediterranean area. Many of these common herbs are found in North Africa, either wild or in gardens. Among the more common of these may be mentioned ANISE, CARAWAY, CELERY, DILL, CUMIN, CORIANDER, FENNEL and PARSLEY, all members of the *Umbelliferae* (Carrot Family) whose seed-like fruits are used for flavoring; and others as FENUGREEK, MAJORAM, THYME, SAGE, BALM, and SAFFRON, and the ubiquitous CHILI PEPPER. Native cookery utilizes many of these and others less well known. It is difficult to estimate the amounts of various herbs grown, for the usual practice is to plant them in small quantities. A few figures are available for Morocco, however, where cumin and coriander are particularly important. In 1936, 1,600 tons of cumin were produced; in 1937, 200 tons; and in 1938, 100 tons. Figures for coriander seed are 6,500; 4,300; and 58. The fruits of coriander are a common flavoring substance for both sweet and savory dishes. When candied in a sugar solution, they are sold as "sugar plums" in Europe. The fruits of cumin are hot and aromatic and are used to flavor various types of food. In Morocco, SPEARMINT (*Mentha spicata*), or a related species, is quite important since mint is always steeped with the green tea of which Moroccans are so fond. The SAFFRON CROCUS is cultivated to some extent. The styles (the upper part of the pistil) of the flower are cut off and dried for use as a spice and yellow dyestuff. Quantities of BLACK PEPPER and other tropical spices were imported, for the Mediterranean climate is quite unsuited to these plants.

Appendix I

SURVEY OF AFRICA'S FOOD RESOURCES

By

Thomas S. Githens

SURVEY OF AFRICAN FOOD RESOURCES

	Nth. Rhodesia	Nyasaland	Tanganyika	Kenya	Uganda	Angola	Belgian Congo	Middle Congo	Gabon	Cameroons	Nigeria	Dahomey	Gold Coast	Ivory Coast	Liberia	Sierra Leone	French Guinea	Gambia	Senegal	Egypt. Sudan	Chad	Niger	French Sudan	Mauritania
Bananas			**i**	**i**	i	i	**ix**		i	**x**	**x**	i	**x**	x	i		x		x	u				
Plantain					**i**		**i**		i	i	i	i	i	i	i	i	i	i	i				i	
Pineapples						i	i		u		i	i	i	x	i		x							
Watermelons											u		i							i	i	i	i	i
Oranges	u		i	i	i	i	i	u	u		i	i			i				i					
Grapefruit	u	ix																						
Limes											x		x		i									
Citron							i															u		
Lemons						u														u				
Grapes						i																		
Mangoes			i	i	i	i	i		u															
Guavas						i	i																	
Tamarinds																								
Pawpaws			i	i	i	i	i																	
Custard apples						i	i																	
Avocadoes							i																	
Date palms											u									ix	i	i	i	i
Oil Palms			i			**x**	**x**	**x**	**x**	**x**	**x**	x	x	x	**x**	**x**	x	**x**	x					
Benniseed			x	**i**	x		ix			i	x		i	x			x		x	**x**	i	i	i	i
Shea Nut										i	x	x	x	x			x		x		i	i	x	
Coconuts			x	i		x			x	x	i	x	x		u				x					

Key:

- x Export crop.
- i Important.
- u Unimportant.
- **x** Boldface is most important crop of its category.
- **i** Boldface is most important crop of its category.

SURVEY OF AFRICAN FOOD RESOURCES (*continued*)

	Nth. Rhodesia	*Nyasaland*	*Tanganyika*	*Kenya*	*Uganda*	*Angola*	*Belgian Congo*	*Middle Congo*	*Gabon*	*Cameroons*	*Nigeria*	*Dahomey*	*Gold Coast*	*Ivory Coast*	*Liberia*	*Sierra Leone*	*French Guinea*	*Gambia*	*Senegal*	*Egypt. Sudan*	*Chad*	*Niger*	*French Sudan*	*Mauritania*
Sugar Cane			x	i	i	x	x				i		i		i									
Cacao			x			x		u	x	x	x	x	x	x	x	x			x		x	x		
Kola							ix			i	i	i	x	x	i	x	x	i						
Coffee	x	x	x	x	x	x	x	x			i		x	x	x	x	i				x	x		
Tea		x	x	x																				
Rice		ix	x	u	u	ix	x			u	i	u	u	i	i	i	i	i	i		u	i	i	i
Maize	i	ix	i	i	i	x	x		i	i	x		i	i	i	i	i	i	i	i		i	i	i
Guineacorn	i	i	i	i	i		i			i	i	i	i	i	i				i	xi		i	i	
Bulrush millet	i	i	i	i	i	i	i			i	i	i	i	i			i		i			i	i	i
Eleusine	i	i			i		i				u													
Finger Millet											i		u			u		u				i		
Maiwa													u											
Fonio											u		i	i			i			i	i	i	i	i
Wheat	i	u	i	x	u	ix	u				u									i		i	u	u
Barley			u																	i		u	u	u
Oats	u		u																					
Peanuts	i	ix	x	x	x	u	x			x	x	x	i	i	u	x	x	x	x	i		x	x	i
Haricot Beans		i				ix	i		i	i		i		i					i			i	i	i
Lima Beans							i			u	i	u	i	u	u	i	u	i		i				
Cow Peas	i										i		i											
Pigeon Peas		i					i				u	i	i	i		i	i							
Sword Beans										u	u	u	u	u	u	u	u	u	u		u	u	u	u
Bambara Nuts							i				i									u	i			
Lab Lab Beans												u										u		

SURVEY OF AFRICAN FOOD RESOURCES (*continued*)

	Nth. Rhodesia	*Nyasaland*	*Tanganyika*	*Kenya*	*Uganda*	*Angola*	*Belgian Congo*	*Middle Congo*	*Gabon*	*Cameroons*	*Nigeria*	*Dahomey*	*Gold Coast*	*Ivory Coast*	*Liberia*	*Sierra Leone*	*French Guinea*	*Gambia*	*Senegal*	*Egypt. Sudan*	*Chad*	*Niger*	*French Sudan*	*Mauritania*
Mungo Beans		i																		i				
Locust Beans											u	i												
Peas					i	u	u																	
Soybeans		i																						
Yams			i			i	u		i	i	i	i	i	i	i	i	i	i		i		i		
Coco-yams			u	u			i			i	i	u	i	i	i	i	i	i						
Cassava	i	i	i	u	i	i	ix		i	i	i	i	i	i	i	i	i	i	i		i	i	i	
Sweet Potato	i	i	i	i	i	i	x		i	i	i	i	i	i	i	i	i	i	i	u	i	i	i	i
Fra-Fra Potato							i			i	u	i	u	i					i			i	i	
White Potato	i	ix	i	i	u	u	ux																	
Ginger							u				i	u	i			x								
Onions						u	i				i									i		i		
Tiger Nuts											u													
Carrots						u	u																	
Okra									u		i	i	i	i	i			i		i				
Eggplant							i		u		u		u			u								
Tomatoes						i	i		u		u		i									i		
Pumpkin			u								u		u							u				
Squash											u													
Cucumbers																				u				
Cucurbits						i														i		i		
Red Pepper		x	u		i	i	ix				i		i		u	x								
Black Pepper											u				u					x				
Guinea Pepper											u			x	i									

Appendix II

FOOD PLANTS OF AFRICA

(Including both native and introduced species)

CEREALS

Barley . . . . *Hordeum vulgare.*
Burgu . . . . *Panicum burgii, Echinochloa stagnina.*
Fonio . . . . *Paspalum longiflorum.*
Guineacorn . . . . *Sorghum vulgare.*
Maiwa . . . . *Pennisetum spicatum.*
Maize . . . . *Zea mays.*
Millet, Bulrush . . . . *Pennisetum typhoideum.*
Millet, Finger . . . . *Digitaria exilis.*
Millet, Finger . . . . *Digitaria iburua.*
Oat . . . . *Avena sativa, Avena* spp.
Rice . . . . *Oryza sativa.*
Rye . . . . *Secale cereale.*
Tamba . . . . *Eleusine corocana.*
Wheat . . . . *Triticum aestivum.*

LEGUMES

Bambara groundnut . . . . *Voandezia subterranea.*
Bean, Broad . . . . *Vicia faba.*
Bean, Haricot . . . . *Phaseolus vulgaris.*
Bean, Lab Lab . . . . *Dolichos lablab.*
Bean, Lima . . . . *Phaseolus lunatus.*
Bean, Locust . . . . *Parkia filicoidea, P. biglobosa.*
Bean, Soy . . . . *Glycine max.*
Bean, Tepary . . . . *Phaseolus acutifolius.*
Bean, Mungo . . . . *Phaseolus mungo, P. aureus.*
Bean, Sword . . . . *Cannavalia ensiformis.*
Carob . . . . *Ceratonia siliqua.*
Chickpea . . . . *Cicer arietinum.*
Geocarpa groundnut . . . . *Kerstingiella geocarpa.*
Lentil . . . . *Lens esculenta.*
Fenugreek . . . . *Trigonella foenum-graecum.*
Peanut . . . . *Arachis hypogea.*
Pea, Chick . . . . *Cicer arietinum.*
Pea, Cow . . . . *Vigna unguiculata, V. catjang.*
Pea, Garden . . . . *Pisum sativum.*
Pea, Pigeon . . . . *Cajanus indicus.*

ROOT CROPS

Artichoke (Jerusalem) . . . *Helianthus tuberosus.*
Carrot . . . . *Daucus carota.*
Cassava, Bitter . . . . *Manihot utilissima.*

Cassava, Sweet . . . . *Manihot palmata.*
Coco-yam (new) . . . . *Xanthosoma mafaffa, X. sagittifolium.*
Coco-yam (old) . . . . *Colocasia esculenta.*
Ginger . . . . *Zingiber officinale.*
Onion . . . . *Allium cepa, A. angolense, Allium* spp.
Potato, Fra-Fra . . . . *Coleus rotundifolius.*
Potato, Sweet . . . . *Ipomoea batatas.*
Potato, White . . . . *Solanum tuberosum.*
Tiger nut . . . . *Cyperus esculentus.*
Yam, Bulb-bearing . . . . . *Dioscorea bulbifera.*
Yam, Chinese . . . . *Dioscorea esculenta.*
Yam, Three-leaved . . . . . *Dioscorea dumetorum.*
Yam, Water . . . . *Dioscorea alata.*
Yam, White . . . . *Dioscorea rotundata.*
Yam, Yellow . . . . *Dioscorea cayennensis.*

VEGETABLES

Artichoke, Globe . . . . *Cynara scolymus.*
Asparagus . . . . *Asparagus officinalis.*
Breadfruit . . . . *Artocarpus incisa.*
Cabbage . . . . *Brassica oleracea.*
Cantaloupe . . . . *Cucumis melo.*
Carrot . . . . *Daucus carota.*
Celery . . . . *Apium graveolens.*
Cucumber . . . . *Cucumis sativus.*
Eggplant . . . . *Solanum melongeana.*
Fennel . . . . *Foeniculum vulgare* var. *dulce.*
Gourd . . . . *Cucurbita pepo.*
Lettuce . . . . *Lactuca sativa.*
Muskmelon . . . . *Cucumis melo.*
Pumpkin . . . . *Cucurbita pepo.*
Okra . . . . *Hibiscus esculentus.*
Spinach . . . . *Basella* sp., *Amaranthus, Rumex* spp.
Squash . . . . *Cucurbita pepo, C. maxima.*
Tomato . . . . *Lycopersicum esculentum.*
Turnip . . . . *Brassica rapa.*
Watermelon . . . . *Citrullus vulgaris.*

FRUITS AND NUTS

Almond . . . . *Prunus communis.*
Apricot . . . . *Prunus armeniaca.*
Avocado . . . . *Persea americana.*
Banana . . . . *Musa sapientum.*
Cherimoya . . . . *Anona cherimoya.*
Cherry . . . . *Prunus* spp.
Chestnut . . . . *Castanea sativa.*
Citron . . . . *Citrus medica.*
Custard apple . . . . *Anona mannii, A. senegalense*
Date . . . . *Phoenix dactylifera.*
Fig . . . . *Ficus carica.*
Grape, Wild . . . . *Vitis* spp.

Grape, Wine . . . . *Vitis vinifera.*
Grapefruit . . . . *Citrus maxima.*
Guava . . . . *Psidium guajava.*
Jujube . . . . *Ziziphus jujuba.*
Kumquat . . . . *Fortunella margarita, F. japonica.*
Lemon . . . . *Citrus limonia.*
Lime . . . . *Citrus aurantifolia.*
Loquat . . . . *Eriobotrya japonica.*
Mandarin . . . . *Citrus nobilis* var. *deliciosa.*
Mango . . . . *Mangifera indica.*
Medlar . . . . *Mespilus germanica.*
Monkey Plum . . . . *Parinarium macrophyllum.*
Orange (sweet) . . . . *Citrus sinensis.*
Orange (bitter) . . . . *Citrus aurantium.*
Pawpaw (Papaya) . . . . . . *Carica papaya.*
Peach . . . . *Prunus persica.*
Pear . . . . *Pyrus communis.*
Pineapple . . . . *Ananas comosus.*
Plantain . . . . *Musa paradisiaca.*
Plum . . . . *Prunus domestica, Prunus* spp.
Pomegranate . . . . *Punica granatum.*
Quince . . . . *Cydonia oblonga.*
Tamarind . . . . *Tamarindus indica.*
Tangerine . . . . *Citrus nobilis* var. *deliciosa.*
Walnut . . . . *Juglans regia.*

CONDIMENTS AND SWEETS

Coriander . . . . *Coriandrum sativum.*
Cumin . . . . *Cuminum cyminum.*
Pepper, black . . . . *Piper nigrum, P. guineense.*
Pepper, cayenne . . . . *Capsicum annuum.*
Pepper, Chilli . . . . *Capsicum frutescens.*
Pepper, Guinea . . . . *Xylopia ethiopica.*
Saffron . . . . *Crocus sativus.*
Sorghum . . . . *Sorghum vulgare* var. *saccharatum.*
Sugar cane . . . . *Saccharum officinarum.*
Vanilla . . . . *Vanilla planifolia.*

OIL PLANTS

Argan . . . . *Argania sideroxylon.*
Benniseed . . . . *Sesamum orientale.*
Coconut . . . . *Cocos nucifera.*
Oil palm . . . . *Elaeis guineensis.*
Olive . . . . *Olea europea.*
Peanut . . . . *Arachis hypogea.*
Shea tree . . . . *Butyrospermum parkii.*

CAFFEIN PLANTS

Cocoa . . . . *Theobroma cacao.*
Coffee . . . . *Coffea arabica, C. robusta, C. liberica.*
Kola . . . . *Cola acuminata, C. nitida.*
Tea . . . . *Thea sinensis.*

BIBLIOGRAPHY

GENERAL

Africa: An Economic Survey of the Colonial Empire, London, 1934.

Barker, W. H., "Trade Communications in Tropical Africa," *Economic Geography*, I (1925), 315.

Broomhall, G. J. S., *Corn Trade Yearbook*, Liverpool, 1935.

Economic Advisory Council, *Nutrition in the Colonial Empire*, London, 1939.

Fitzgerald, W., *Africa: Social, Economic and Political Geography*, 4th ed., New York, 1942.

Gouvernement général de l'Afrique occidentale française, *Les grands produits de l'Afrique occidentale française*, Paris, 1931.

Gurney, H. F., *Economic and Commercial Conditions in the British Dependencies of West Africa in 1936*, London, 1937.

Hailey, M., *An African Survey*, London, 1938.

Herbertson, A. J., and Howarth, M. A., *Oxford Survey of the British Empire*, Vol. II, Oxford, 1914.

International Institute of Agriculture, *International Year Book of Agricultural Statistics*, Rome, 1938–39.

International Institute of Agriculture, *International Review of Agriculture*, Rome, 1938, 1939, 1940.

L'Afrique française, Bulletin mensuel du Comité de l'Afrique française et du Comité du Maroc, Paris, 1931.

League of Nations Statistical Year Book, Geneva, 1938, 1939, 1940.

Shantz, H. L., "Agricultural Regions of Africa," *Economic Geography*, XVI (1940), 1, 122, 341; XVII (1941), 217, 353.

Shantz, H. L., and Marbut, C. F., *The Vegetation and Soils of Africa*, New York, 1923.

WEST AFRICA

Abadie, M., *La Colonie du Niger*, Paris, 1927.

Agence économique de l'Afrique occidentale française, *Bulletin mensuel*, XII, Toulouse, 1931.

Cameroons. *Colonial Office Report for 1938.*

Cummins, H., "Botany of the Ashanti Expedition," *Bulletin of Miscellaneous Information*, Royal Gardens, Kew, April 1898.

Faulkner, O. T., and Mackie, J. R., *West African Agriculture*, Cambridge, 1933.

Francois, M. G., *Les Productions de l'Afrique occidentale française*, Melun, 1918.
Gambia, *Blue Book for 1936*, London, 1937.
Gambia, *Colonial Office Report for 1938*, London, 1938.
Gold Coast, Department of Agriculture, *Reports 1935–41*.
Department of Agriculture, *Year Book for 1928–29*.
Gouvernement general de l'Afrique occidentale française, *Le Soudan français*, Paris, 1931.
Irvine, F. R., *West African Agriculture, Soils, and Crops*, London, 1934.
Nigeria, Agricultural Department, *Annual Reports 1930–1938*, Lagos, 1931–39.
Annual Report on the Northern Provinces for 1936, Kaduna, 1937.
Annual Report on the Southern Provinces for 1935, Lagos, 1937.
Annual Report on the Social and Economic Progress for 1937, Lagos, 1938.
Blue Book of Statistics, 1930–1938, Lagos, 1931–39.
Colonial Office Reports, 1930–1938, London, 1931–39.
Reading, J. H., *Newest Africa: A Voyage Along the Western Coast*, London, 1914.
Renner, G. T., "Geographic Regions of the (French) Sudan," *Economic Geography*, II (1936), 256.
Roest, P. K., "French Equatorial Africa," *Foreign Agriculture*, VI (1943), 3. United States Department of Agriculture.
Senegal, *British Colonial Office Report for 1938*, London, 1938.
Sierra Leone, *Blue Book for 1938*, London, 1938.
Colonial Office Report for 1938, London, 1939.
Strong, R. P., et al., *The African Republic of Liberia*, Cambridge, 1930.
Togoland, *Colonial Office Report for 1938*, London, 1939.
Vuillet, J., "L'Agriculture dans le pays de Segou (Niger)," *L'Afrique française*, XXX (1920), Supplement, 169.

WEST EQUATORIAL AFRICA

British Consular Reports, #5529, Angola, London, 1915.
British Consular Reports, #5497, South Angola, London, 1915.
Bullock, G. H., *Economic Conditions in Angola*, Great Britain, Department of Overseas Trade, London, 1934.
Goodstein, S. S., "Angola," *Foreign Agriculture*, VII (1943), 5.
Hutcheson, A. B., *Report on Economic Conditions in Angola*, London, 1923.
Manual of the Belgian Congo, London, 1921.
O'Meara, F., *Report on the Economic and Commercial Conditions in Angola*, London, 1937.

Publications de l'État indépendant du Congo, Paris, 1893.
Statistiques sur le commerce extérieur du Congo Belge, 1937, Brussels, 1937.
United States Department of Commerce, *Trade Information Bulletin, # 254, Angola*, Washington, D. C., 1924.
Year Book of Compared Colonial Documentation, 1937, *Belgian Congo*, Brussels, 1937.

NORTH EAST AFRICA

Abyssinia, *British Diplomatic and Consular Reports*, London, 1910–15.
United States of America, Department of Commerce, *Consular Report # 81*, Washington, D. C., 1918.
Anglo-Egyptian Sudan, Handbooks, 1. *Bahr el Ghazal*, London, 1911; 2. *Kordofan and White Nile*, London, 1911.
Report on Finances, Administration and Condition of the Sudan in 1931, London, 1931.
Report on the Condition of the Sudan in 1930, London, 1931.
United States of America, Department of Commerce, *Trade Promotion Series*, # 49, Washington, D. C., 1927.
Gleichen, Count, *The Anglo-Egyptian Sudan*, London, 1905.
Somaliland, *Colonial Office Reports for 1938–1939*, London, 1939.
La Côte française de Somali, Melun, 1919.

EAST AFRICA

Brigden, J. W., *Economic Conditions in Northern Rhodesia and Nyasaland for 1934–1935*, London, 1934, 1935.
British East Africa, *British East Africa: Kenya, Uganda, Tanganyika, and Somaliland*, London, 1903.
Conquered East Africa and Its Resources, Handbook # 85, London, 1918.
Department of Agriculture for British East Africa, *Annual Reports*, Nairobi, 1917–18.
Imperial Economic Committee, Report # 28, *Maize*, London, 1934.
United States of America, Department of Commerce, Trade Information Bulletin # 173, *British East Africa*, Washington, D. C., 1924.
Brode, H., *British and German East Africa*, New York, 1911.
Buchanan, J., *The Shire Highlands*, London, 1885.
Fitzgerald, W., *Travels in British East Africa*, London, 1898.
Gouldsbury, G., and Sheane, H., *The Great Plateau of Northern Rhodesia*, London, 1911.
Johnston, Sir H., *The Uganda Protectorate*, New York, 1902.

Kenya, *Blue Book for 1937*, Nairobi, 1938.
Colonial Office Reports, 1930–1938, London, 1931–39.
Native Affairs Department, *Annual Report for 1932*, London, 1934.
Report of the Agricultural Commission for 1929, Nairobi, 1929.
Lamb, P. H., "Recent Agricultural Developments in Uganda," *Bulletin of the Imperial Institute*, X (1912), 422.
Mitchell, N. P., *Land Problems and Policies in the African Mandates of the British Commonwealth*, Baton Rouge, 1931.
Northern Rhodesia, *Colonial Office Reports, 1930–1938*, London, 1930–38.
Nyasaland, *Blue Book for 1937*, Zomba, 1938.
Colonial Office Report for 1938, London, 1939.
Department of Agriculture, *Annual Report*, Zomba, 1927, 1932.
General Information on the Nyasaland Protectorate, London, 1920.
Handbook, Zomba, 1932.
Report of the British Central African Protectorate, London, 1907.
Tanganyika, *Annual Report to the League of Nations*, London, 1937, 1938.
Blue Book for 1937, Dar es Salaam, 1937.
Tanganyika Guide, Dar es Salaam, 1936.
Tanganyika Territory Trade Reports, Dar es Salaam, 1938.
United States of America, Department of Commerce, *Trade Information Bulletin*, #86, Washington, D. C., 1923.
Uganda, *Bluebook for 1938*, Entebbe, 1939.
Colonial Office Reports, 1915–1918, 1930–1938, London, 1915–18, 1930–38.
Wallis, H. R., *Handbook of Uganda*, London, 1913.

NORTH AFRICA

General

Bernard, Augustin, *L'Afrique septentrionale et occidentale*, Pt. I: Vidal de la Blache, P., and Gallois, L., *La Géographie universelle*, XI, Pt. I, Paris, 1937.
Bernard, Augustin, and de Flotte de Roquevaire, R., *Atlas d'Algérie et de Tunisie*, Fasc. XIII, "Cartes Agricoles," Paris-Algiers, 1933.
Clévénot, Elisabeth, *La Production des Primeurs dans l'Afrique du Nord et leur Commerce d'exportation: Etude géographique et économique*, Algiers, 1932.
Fairchild, David, *Exploring for Plants*, New York, 1931.
Foreign Commerce Weekly, "Oils, Fats and Oil Seeds," IX (1942), 27–28.
Gautier, E. F., *Sahara, the Great Desert*, translated by Dorothy Ford Mayhew, New York, 1935.

——, *L'Afrique blanche*, Paris, 1939.

Gottman, Jean, "Economic Problems of French North Africa," *Geographical Review*, XXXIII (1943), 175–96.

Hardy, Georges, and Ricket (eds.), *L'Alimentation Indigène dans les Colonies françaises*, Paris, 1933.

International Institute of Agriculture, "L'Olivier dans le Monde: Superficie, Production, Commerce de ses Produits," *Série de monographies sur les principaux produits agricoles du marché mondiale*, no. 6, Rome, 1939.

Pascual, A., "Almond Growing Throughout the World," *International Review of Agriculture*, XXXI (1940), 113–48T, International Institute of Agriculture.

——, "Citrus Production in Morocco and Tunisia," *International Review of Agriculture*, XXXI (1940), 425–31T, International Institute of Agriculture.

Algeria

Bernard, Augustin, *L'Algérie*, Paris, 1931.

Encyclopédie coloniale et maritime, Series II: *L'Algérie*, Fasc. 1, "L'Algérie—Monographie générale," Paris, 1936.

——, Series II, Fasc. 6–7, "L'Algérie Agricole," Paris, 1936.

Forbin, Victor, "L'Industrie vinicole en Algérie," *La Nature*, LXII (1934), 481–89.

Pascual, A., "Citrus Growing in Algeria," *International Review of Agriculture*, XXXI (1940), 41–47, International Institute of Agriculture.

Taylor, Griffith, "Sea to Sahara: Settlement Zones in Eastern Algeria," *Geographical Review*, XXIX (1939), 177–95.

Morocco

Bernard, Augustin, *Le Maroc*, Paris, 1932.

——, Series IV, Fasc. 2–3, Jean Célérier, "Les grandes regions naturelles du Maroc, Paris, 1938.

Encyclopédie coloniale et *maritime*, Series IV, *Maroc*, Fasc. 6–7, Gaston Lebault, "Le Maroc agricole," Paris, 1938.

Hardy, Georges, "L'Alimentation des Indigènes au Maroc": Hardy and Ricket, *L'Alimentation indigène dans les Colonies françaises*, Paris, 1933.

Maroc, *Atlas historique*, *géographique* et *économique*, Paris, 1935.

Miege, E., "La question du sucre et de la betterave au Maroc," *Revue de Geographie Marocaine*, XXIV (1940), 11–52.

Sermaye, Jean, "La politique de l'eau au Maroc francais," *L'Illustration*, CCI (1938), 249–52.

Société d'Etudes économiques et statistiques, *Bulletin économique du Maroc*, I–VI (1934–39).

Tunisia

Bena, Henri, "The Progress of Oil Cultivation and Manufacture in Tunisia," *International Review of Agriculture*, XXXI (1940), 147–52E, International Institute of Agriculture.

Encyclopédie coloniale et maritime, Series III, *Tunisie*, Fasc. 1, H. Bec, J. Gasser, H. Laverdet, "La Tunisie française—Monographie générale," Paris, 1936.

——, Series III, Fasc. 4, F. Boeuf, R. They, Ed. Coanet, Ch. Chedeville, "La Tunisie économique—Les Céréales, l'Olivier, la Viticulture," Paris, 1937.

——, Series III, Fasc. 5, H. Rebour, M. Lescuyer, H. Cottier, "La Tunisie économique—l'Arboriculture, les Forêts, l'Elévage," Paris, 1937.

Gomer, "La Tunisie agricole et la crise," *Outre Mer*, V (1933), 141–46.

Tunis, Service Botanique, *Annales*.

——, Direction de l'Agriculture et du Commerce, *Bulletin*.

La Tunisie Agricole, XXXVI–XLI (1935–41).